Cognitive Robotics

Cognitive Robotics

edited by

Hooman Samani

Department of Electrical Engineering
National Taipei University
Taiwan
www.hoomansamani.com

CRC Press
Taylor & Francis Group
Boca Raton London New York

CRC Press is an imprint of the
Taylor & Francis Group, an **informa** business

CRC Press
Taylor & Francis Group
6000 Broken Sound Parkway NW, Suite 300
Boca Raton, FL 33487-2742

Printed on acid-free paper
Version Date: 20150814

International Standard Book Number-13: 978-1-4822-5403-7 (Hardback)

Library of Congress Cataloging-in-Publication Data

Therapeutic medicinal plants from lab to the market / editors: Marta Cristina Teixeira Duarte,
   Mahendra Rai.
     pages cm
   Includes bibliographical references and index.
   ISBN 978-1-4822-5403-7 (hardcover : alk. paper) 1. Materia medica, Vegetable. 2. Medicinal
   plants. 3. Botanical drug industry. I. Duarte, Marta Cristina Teixeira, editor. II. Rai, Mahendra,
   editor.

RS164.T52 2016
615.3'21--dc23
                                                            2015028938

Visit the Taylor & Francis Web site at
http://www.taylorandfrancis.com

and the CRC Press Web site at
http://www.crcpress.com

Contents

Section VI Psychological Aspect of Cognitive Robotics

CHAPTER 9 ■ On the Crossroads of Cognitive Psychology and Cognitive Robotics   171

ROY DE KLEIJN, GEORGE KACHERGIS, and BERNHARD HOMMEL

Section VII **Artificial Intelligence Aspect of Cognitive Robotics**

Contributors

Paolo Barattini
Ridgeback SAS
Cuneo, Italy

Na Chen
Department of Industrial
Engineering Tsinghua University
Beijing, China

Alex Yu-Hung Chien
Department of Computer Science
National Tsing Hua University
Hsinchu City, Taiwan

Belinda J. Dunstan
Creative Robotics Lab University of
New South Wales Art and Design
Sydney, Australia

Bernhard Hommel
Leiden Institute for Brain and
Cognition Leiden University
Leiden, Netherlands

George Kachergis
Leiden Institute for Brain and
Cognition Leiden University
Leiden, Netherlands

Alexis Karkotis
BornAnIdeaLab
London, United Kingdom

Roy de Kleijn
Leiden Institute for Brain and
Cognition Leiden University
Leiden, Netherlands

Jeffrey Tzu Kwan Valino Koh
Creative Robotics Lab University of
New South Wales Art and Design
Sydney, Australia

Christian Ø. Laursen
Aarhus University
Aarhus, Denmark

Jürgen Leitner
Dalle Molle Institute for AI (IDSIA)
and SMRTRobots, Lugano,
Switzerland

David Levy
Intelligent Toys Ltd.
London, United Kingdom

Tim R. Merritt
Aarhus School of Architecture
Aarhus, Denmark

Mie Nørgaard
MIE NØRGAARD
Værløse, Denmark

Marianne G. Petersen
Aarhus University
Aarhus, Denmark

Doros Polydorou
University of Hertfordshire
Hatfield, United Kingdom

Majken K. Rasmussen
Aarhus University
Aarhus, Denmark

Pei-Luen Patrick Rau
Department of Industrial
Engineering Tsinghua University
Beijing, China

Von-Wun Soo
Department of Computer Science
National Tsing Hua University
Hsinchu City, Taiwan

Hans Sprong
Philips RoboCup
Eindhoven, Netherlands

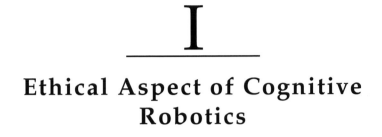

I

Ethical Aspect of Cognitive Robotics

CHAPTER 1

When Robots Do Wrong

David Levy

Intelligent Toys Ltd., London, United Kingdom.

CONTENTS

ALREADY mankind has begun to embrace the "robot society." In 2005 the government of South Korea announced its intention to have a robot in every household by 2020. And the Japanese Robot Association predicts that Next Generation Robots will generate up to $ 65 billion of economic activity by 2025. Also in Japan efforts have been under way for some time to ensure that, before long, elderly members of the population will routinely have robots to take care of them. Meanwhile, at many universities and other research centres, robots of just about every flavour are a hot topic.

1.1 INTRODUCTION

Clearly robots will soon be assisting us in many different aspects of our lives, becoming our partners in various practical and companionable ways and entertaining us. An early example in the field of entertainment was a ballroom dancer robot that was unveiled in 2005 at the World Expo in Japan. It did not have usable legs but instead moved on three wheels.

Let us consider the following scenario, perhaps ten years from now, when dancing robots do have moveable legs and possess the skills for a variety of dance steps, such as waltz, foxtrot, rumba, ... One evening a young lady named Laura is at a dance, partnering such a robot. The band strikes up with the music for a cha cha cha but performs it so badly that the robot mistakes the music for a tango. So Laura and her robot partner are holding each other but dancing at cross purposes, and very soon they fall over. The robot lands on top of Laura and, being quite heavy, breaks both of her legs. Laura's father is furious, and calls his lawyers, telling them to commence legal proceedings immediately and "throw the book at them."

But at whom should his lawyers throw the book? Should it be the dance hall, or the online store that sold the robot to the dance hall, or the manufacturer of the robot, or the robot's designers, or should it be the independent software house that programmed the robot's tune recognition software, or the band leader, or even the whole band of musicians for playing the cha cha cha so badly?

Some months later the trial opens with all of these as defendants. Expert witnesses are called - experts on everything from robot software to motion sensor engineers to the principals of dance music academies. What do you think would be the result of the trial? I'll tell you - the lawyers do very nicely thank you.

As technology advances, the proliferation of robots and their applications will take place in parallel with increases in their complexity. One of the disadvantages of those increases will be a corresponding increase in the number of robot accidents and wrongdoings. How will legal systems be able to cope with all the resulting court cases? In fact, will they be able to cope? The answer, surely, is "No."

In this chapter, I am going to addresses the questions: What should happen when something goes wrong? Who, or what, is responsible? And above all, how best should society deal with an ever-increasing number of robot accidents. All developed countries will be faced with these problems, and all countries need to adopt legally and ethically sound approaches to finding solutions to them.

1.2 BLAME THE ROBOT?

Even though a robot is a manmade object, one superficially reasonable way to apportion blame for a robot accident or wrongdoing would be to blame the robot itself. Many people support this idea because robots are autonomous.

I shall discuss three possible approaches to this particular argument:

a. A robot should be regarded as a quasi-person;

b. A robot should be regarded as a quasi-animal; and

c. A robot should be regarded as neither, but simply as a product, a manmade object, period.

1.2.1 The Robot as a Quasi-Person

As computer software, and therefore robots, gain in intelligence, many of us will come to accept that their mental capabilities have a humanlike quality. Leon Wein argues that a robot's capacity to make decisions and act on them enables them to:

> "operate in a manner identical to that of humans ... society may have no choice but to accept machines as "legal persons" with rights as well as duties."

These duties will include the obligation to act in a manner that would be considered reasonable if it were human behaviour. The same obligation applies to corporations, which, despite not being human, have legal rights and responsibilities and can be punished for their transgressions and their negligence. This possibility, punishing corporations, points the way to the concept of punishing robots, by fining them for example.

1.2.2 The Robot as a Quasi-Animal

Some researchers argue that because robots, like animals, are autonomous, society can justify blaming robots for their accidents.

For the purposes of attributing legal responsibility for accidents and wrongdoings, the analogy of robots as domesticated animals is a concept that has gained a fair measure of support in the recent literature. Richard Kelley and his colleagues at the University of Nevada draw an analogy between the different breeds of dogs and the different functions performed by robots. For example, a Roomba vacuum cleaner robot is analogous to a breed of dog that is largely harmless, but an unmanned aircraft drone, especially if it is armed, is analogous to a dangerous breed. And for "dangerous" robots Kelly and his colleagues suggest bans or restrictions similar to those imposed on the owners of dangerous dogs. But for robots that are considered "safe," such as small toys or service robots, they do not advocate any such restrictions. Based on these analogies they propose having different laws pertaining to different types of robot, extending to serious criminal penalties in some cases for the owner of a "dangerous" robot that causes damage.

In his classic 1981 paper "Frankenstein Unbound," Sam Lehman-Wilzig suggests that the punishment of dangerous robots and their owners could be achieved by "rehabilitating" the robot, in other words reprogramming the culprit. He also proposes, for less severe offences, putting a robot to work for the benefit of its human victim(s), in order to make restitution for the robot's wrongdoing. Lehman-Wilzig has pointed out that, as they grow more intelligent, the level of damage that robots can inflict becomes greater, so perhaps the onus of responsibility for their accidents should shift to their owners or end-users. On this basis he argues that the principles applied to dangerous animals should also be applied to robots. However, I believe that, no matter how intelligent an animal might appear to be, it should always be the owner (or handler) rather than the animal itself who should be held culpable for any accident or mischief that the animal causes.

1.2.3   The Robot as a Manmade Object

Not everyone will come to accept the concept of robots as quasi-persons or quasi-animals, leaving us to regard robots as nothing more than manmade objects.

From a legal perspective the relationship between owner and robot is known as "agency" — the owner acts by means of (or through) their robot, because the robot's actions are generally commanded by the owner/operator. The legal principle of agency is that *"he who acts through another is deemed in law to do it himself"*, and therefore, under some circumstances, it is the owner/operator who, through their agent (the robot), is legally responsible for a robot accident or wrongdoing.

1.2.4   Punishing the Robot

Those who believe that it makes sense to blame the robot itself for its accidents and wrongs, need to consider how the robot might be punished. This is hardly a novel question. In 1985 Robert Freitas Jr. discussed some of the complications surrounding robot punishment in his seminal paper "The Legal Rights of Robots:"

"Robots could be instantly reprogrammed, perhaps loading and running a new software applications package every hour. Consider a robot who commits a felony while running the aggressive "Personality A" program, but is running mild-mannered "Personality M" when collared by the police. Is this a false arrest? Following conviction, are all existing copies of the criminal software package guilty too, and must they suffer the same punishment? (Guilt by association?) If not, is it double jeopardy to take another copy to trial?"

And to Freitas's comments Yorick Wilks added an even more drastic solution:

"But what can we say of "machine punishment"? A machine can be turned off and smashed and the software will either go with it, or can be burned separately, provided we know we have all the copies!"

If we fast forward a quarter of a century from Freitas's paper, to the year 2010, we find Gabriel Hallevy's more detailed examination of the concept of punishment for an artificially intelligent entity, within the context of criminal law. Hallevy evaluates five different types of punishment typically imposed on the guilty in human criminal courts, and he considers the applicability of each type to AI entities. These punishments are: the death penalty, imprisonment, a suspended prison sentence, a fine, and community service.

Clearly the death penalty is the most effective method of preventing an offender from reoffending, and the corresponding practical action for achieving an equivalent result with robots is the deletion of its software, thereby rendering the offending robot incapable of committing any further offences. This assumes of course that we accept, as being a different robot, a new robot "brain", i.e. its software, in a pre-existing robot body. Failing such acceptance society would need to rely on Wilks's solution and destroy all copies.

What of imprisonment? The practical action which, for robots, corresponds most closely to imprisonment, is to disable it for a specified period. But passing a prison sentence on a robot, or a suspended sentence, would only have an effect if the robot has been programmed to *dislike* the idea of being deactivated (*imprisoned*).

If a robot is deemed to own money or some other form of property, imposing a fine on it has an analogy in fining a corporation, because neither is human. So when robots are able to earn money through their "work" the idea of a robot being fined will not seem to be at all outlandish, but just as reasonable as fining a corporation.

Community service is another form of punishment which for humans requires the offender to contribute their labour to the community. As Hallevy points out, an AI entity will be able to take on "all sorts of jobs," and therefore:

> "The significance of community service is identical, whether imposed on humans or AI entities."

From Hallevy's analysis it can be argued that most of the types of punishment commonly applied in our courts to human offenders are also applicable, albeit sometimes in different forms, to robots, because the nature of these penalties remains the same relative to humans and to robots. But Peter Asaro queries the whole concept of punishment for robots, asking the questions: Can punishing robots be viewed as a deterrent? Can it achieve justice? And in discussing the relevance of physical punishment for robots Asaro comments that:

> "The various forms of corporal punishment presuppose additional desires and fears of being human that may not readily apply to robots as pain, freedom of movement, mortality, etc. Thus, torture, imprisonment and destruction are not likely to be effective in achieving justice, reform or deterrence in robots. There may be a policy to destroy any robots that do harm but, as is the case with animals that harm people, it would be a preventative measure to avoid future harms rather than a true punishment."

In addition to Asaro's arguments concerning the problematic nature of punishment for robots, I can see two *additional* levels of complexity in the concept. One of these is the matter of deterrence. If a robot is to be deterred by a fear of some form of punishment it must be programmed to recognize that it is contemplating an action regarded by society as being contrary to civil or criminal law. And, it must also be programmed with some sort of artificial emotionally-driven impetus to "dislike" the corresponding punishment for that offence.

The other additional level of complexity that I believe deserves serious consideration, concerns how to deal with robots that are identical to one that has committed an offense. If the errant robot misbehaved because of faulty hardware or the corruption of part of its software, we can argue that all bug-free copies of exactly the same type, model and version of the robot, should be immune from any punishment imposed for a specific case of an offence committed by the buggy errant robot. But if the fault lies in the hardware design or in the programming of the software, then it would seem logical to punish all copies of that robot, since society would be invoking sanctions against the design or programming itself rather than a particular copy of it.

1.2.5 Blaming the Robot — Summary

If blaming the robot itself is to be justified on the grounds that robots are quasi-persons or quasi-animals, it logically follows that the robot itself should be punished, but I find Asaro's arguments against the merit of punishing robots to be rather convincing. And if robots are regarded as nothing more than manmade objects, then the question of how to attribute blame for a wrong will take us into the realm of product liability. Is the product (the robot) defective? And if so, is it fair to blame the robot at all? Human defendants are sometimes able to escape blame based on the defence that they were of "unsound mind" at the moment of their offence, so for robots with design or manufacturing faults the answer must surely be "No". It is not fair. Justice will *not* be served by blaming the robot.

So all things considered, I do not believe that blaming the robot is a sustainable approach to the problem of how society should react to a robot accident.

1.3 BLAMING HOMO SAPIENS

If robots are not to blame for causing their accidents, the blame must therefore fall on we humans: sometimes just one person, sometimes a few individuals, sometimes collectively as is the case with corporations, hospitals, factories, or whatever other body might be responsible in some way for the robot in question. And the blame will either be due to product defects — what the law refers to as product liability - or to product misuse, otherwise known as negligence.

1.3.1 A Question of Trust

Before I discuss the liability and negligence laws as they might be applied to the use of robots, I want to ask the question — how should an owner/operator of a robot reasonably expect their robot to behave and perform? Knowing the answer to this question will help a court to determine, for example, if a robot wrongdoing came about because the robot was faulty, or because its operator incorrectly expected the robot to perform differently from the way in which it was designed to do, or because the operator somehow used or instructed the robot incorrectly when he should have known better.

The reasonable expectation of how a robot will behave and perform is founded on a combination of knowledge and trust. A user's knowledge of what to expect from their robot can simply be found by reading the operating manual. And trusting a robot will develop from first hand and anecdotal experience of using that particular type of robot, or even exactly that particular robot.

To what extent can we safely rely on robots and trust them? Everyone who has flown in a modern passenger aircraft has put their lives in the hands of a robot — an automatic pilot. And the reliability of autopilots is so great that American courts have held human pilots to be negligent for failing to use their autopilots in critical situations. In one case a pilot failed to engage the robot during the approach to an airfield when a state of turbulence existed because of the wake of another aircraft. His plane crashed and the court held that his failure to use his autopilot was a contributing cause.

Another field in which trust is a hugely important element is medicine, where there are many computer and electronic systems employed by doctors to enable them to provide improved care to their patients. The medical robots of the future will have the capability not only of monitoring a patient's vital signs and warning the medical staff when appropriate, but also providing care to the patient through diagnosis, devising drug regimes, and even fully automated robot surgery.

Just how far should we trust such systems? As long ago as the 1970s, a computer program called MYCIN was developed at Stanford University to identify bacteria causing severe infections such as meningitis, and to recommend suitable regimes of antibiotics. An evaluation of MYCIN in comparison with five specialists from Stanford Medical School[1] found that the computer program's "acceptability" performance was rated at 65%, which was clearly superior to that of the five human specialists whose ratings ranged from 42.5% to 62.5%. Who would *you* prefer to trust with *your* health? A human expert, but one who is inferior in performance to MYCIN, or the computer program which performed better than any of the Stanford experts who took part in the trial?

The obvious conclusion for us to draw from these examples is that, for important decision-making tasks such as medical diagnosis, we ought to place more trust in robots and computer software than we currently do. But in fact MYCIN was never actually used in practice. Why on earth not? One reason was the legal objections raised against the use of computers in medicine, asking who should be held responsible if the program were to proffer a wrong diagnosis or to recommend the wrong combination or dosage of drugs.

Another domain in which trust in a robot will be an essential ingredient of the user experience is personal relationships, particularly intimate ones, as social robots increasingly become emotionally appealing to humans. In the coming decades the sophistication of Artificial Intelligence will reach the point where humans falling in love with robots and having sex with them become everyday occurrences. A crucial consideration in human-human intimate relationships will then be fidelity, but not as it is now — a question of whether or not one's spouse or partner is also intimately involved with someone else — but whether they are involved with a sexually capable robot. Last year the American lawyer Sonya Ziaja published a ground-breaking investigation into the legal implications when the consequences of a robot

---

[1] All of whom were faculty members in the Division of Infections Diseases in the departments of Medicine and Pediatrics.

wrongdoing are emotionally rather than materially or physically damaging. In the USA such cases are called "heart balm" torts — they are for alienation of affection against a third party adult who "steals" the affections of the plaintiff's spouse, which is still a civil law offence in eight American states. Ziaja's paper explores the realm of legal liability when a sexually empowered robot is claimed by a marital partner to be at least partly responsible for the destruction of their relationship with their spouse.

In those states of the USA where heart balm torts are still on the statute books, the ethical justification usually claimed for their existence is to protect marriages from breaking up because of third parties. Ziaja posits robots as possible third parties in this equation, and asks:

"If a heart balm suit were to be brought by a spouse where the third party was a sexbot instead of another human, who would be liable? The heart balm torts are directed against the third party paramour. In this hypothetical though, the lover is not human, nor a legal person. When presented with a case like this, legal systems will have to respond by either finding a human at fault (the inventor, manufacturer, the owner) or by finding the sexbot at fault. Either of these options is problematic."

1.4   NEGLIGENCE AND PRODUCT LIABILITY LAWS

The question of legal responsibility relating to robots is highly complex. One aspect of this problem is that in some jurisdictions there is a big legal difference between a robot that goes wrong and a person or persons being negligent with respect to the use of a robot. When a robot malfunctions the applicable legal rules are generally those relating to product liability — is the product well designed, well made, and reliably fit for its purpose? But when a human *uses* a robot it will often be the rules pertaining to human negligence that are relevant. To make matters even more complicated there is also a third type of situation - where it is difficult or even impossible to determine whether the human user of a product is at fault, or whether the fault lies with the product itself, or it could be a combination of both because a service provider will often be using a product to help provide their service. Next is a historically important example.

1.4.1   The First Robot Fatality

In Karel Capek's play *Rossum's Universal Robots* (R.U.R.), which premiered in 1921, the robots take over the world and dispose of the unwanted humans. January 25th 1979 was the 58th anniversary of the first performance of the play, and it was also the date of the first known fatal robot accident, when a factory worker, Robert Williams, was killed at a Ford Motor Company factory in Michigan. Williams died instantly when he was hit on the head by a robot designed by a company called Litton Industries Inc.

Robert Williams's family sued for damages in a jury trial that was tasked with deciding on the issues of negligence and liability. Litton's lawyers claimed that the robot had not been designed to operate when any humans were within its reach. The purpose of the robot was to retrieve parts from their storage locations and then move those parts to an assembly line, and before anyone went near the robot it should first have been switched off so that it couldn't move. Although the robot had a continuously rotating beacon light to warn people when it was switched on and might move at any time, Litton argued that the reason there was no specific warning system to indicate when the robot was actually about to move was because no one was supposed to go near it unless the power was off — a safety rule that Litton claimed was regularly broken by Robert Williams and his co-workers.

The lawyers for Williams's family argued that the design of the robot did not take into account the real-world environment in which the robot was used. In particular, they presented evidence to show that there were many circumstances in which, for reasons of practical necessity, workers might need to be within the area of the robot's operation and reach. For example, a part might be stored in the wrong location in which case the robot would deliver the wrong part to the assembly line, whereupon a human presence was needed to rectify the problem. Because of such problems, Williams's family claimed that the robot's design deficiencies outweighed the fact that Williams and his co-workers would routinely enter the robot's work area while it was moving, rather than cost the factory money by sacrificing production when the robot was switched off.

The trial lasted eight days after which the jury took only a few hours to rule in favour of Robert Williams's family and award them $10 million in damages, having decided that the manufacturer was responsible. But juries do not always make the best decisions, and it is therefore reasonable to ask ourselves, despite that particular jury's decision, who, really should have been held liable for Robert Williams's death? Should it perhaps be the owners of the factory where the accident happened, for not ensuring that adequate safety measures were in place in the factory itself? And what about those who were responsible for testing the robot in the Ford factory to ensure that its use was perfectly safe?

The details of the Williams case presented here provide a salutary example of just how complex a single robot accident case can be. Robot accidents are not just like any other machine accident; robots are not just like any other machine; robots have intelligence, albeit artificial, and being able to "think" places them in a special category that appears to open up all sorts of options as to how we should deal with robot accidents. The fact that there are so many options for attributing blame is surely a convincing argument for trying to find a new approach as starting afresh in the search for a sound and consistent way to compensate the victims of robot accidents.

1.4.2   Robots and the Laws of Negligence

Negligence in a legal sense occurs where there is a duty of care and someone suffers harm if that duty is not properly discharged. The level of care expected is that which a "reasonable" person would apply in the circumstances under which the harm occurred. One of the complications in establishing such a breach is that different standards can reasonably be expected of different defendants, for example doctors are held to a higher standard of medical care than are nurses.

A further element necessary to establish negligence in a court case is something called "causation" — did the defendant cause the harm to the plaintiff? Peter Asaro gives us a succinct explanation:

"Legally culpable forms of negligence depend upon either failures to warn, or failures to take proper care. A failure to warn occurs when the manufacturer was knowingly aware of a risk or danger but failed to notify consumers of this risk. This is the reason why there are now so many warning labels on various products

. . . . . A failure to take proper care or avoid foreseeable risks is more difficult to prove in court because it is more abstract, and involves cases where the manufacturer cannot be shown to have known about a risk or danger from the product. In these cases, it is argued that the given danger or risk was in some sense obvious or easily foreseeable, even if the manufacturer failed to recognize it. In order to prove this, lawyers often bring in experts to testify that those risks were obvious, and so forth. "

While the *concept* of legal negligence is not very difficult to understand, it will be no easy task for the courts to determine negligence in robot cases. Samir Chopra and Laurence White, in their book "A Legal Theory for Autonomous Artificial Agents," provide us with examples of arguments, relating to robots, that could open the flood gates to lawsuits for negligence, including these:

1 Robots can be handled carelessly or dangerously.

2 An operator may not take proper care in programming or configuring a robot before making it available to users.

3 If the operator makes a robot available to a user who is not adequately informed how to use the robot correctly, this can theoretically give rise to liability.

4 There can be a liability for negligent supervision by custodians of decision-making robots.

Clearly one of the ripest fields for robot negligence cases, especially in the USA, is medicine. If a surgical robot makes a wrong incision during an operation, with serious or even fatal consequences, on whom should fall the legal liability for negligence? As the use of medical robot systems becomes even more widespread, there will doubtless be a corresponding increase, particularly in the litigious climate that prevails in the USA, in the number of medical malpractice suits arising from their use. A surgeon's lawyers would likely attribute any blame to the robot's manufacturer on the basis that the robot was defective, while the manufacturer and the hospital would probably blame each other, and the manufacturer's lawyers would most likely also point at the surgeon. Even more fat fees for *all* the lawyers.

1.4.3   Robots and the Laws of Product Liability

Product liability laws are designed to protect people from risks associated with defective products, those that can cause physical injury or some other form of damage. In the USA product liability laws sometimes vary from one state to another, but in general a product's manufacturer is subject to liability for any physical harm caused by the product's use. But matters of product liability are often are not that straightforward. Robots are a special category of product because of their (artificial) intelligence which, in many cases, will include an ability to learn.

Although products should be designed and manufactured in such a way that they will not pose a risk to their users or to others, how could we ever be certain that there will be no risk to people when robots are able to learn from their own experience and hence to improve their own decisions and actions? And if the software in a robot can learn or otherwise change through self-modification, it will often be extremely difficult, if not impossible, to trace the origins of a defect back to the programmer or manufacturer responsible, because it could be argued that the original design of the robot had been altered by the robot itself. In a court it would be necessary to introduce evidence from technical experts as to whether harm arose because of an error in design or manufacture, or whether it was because the robot had been badly trained.

As robots become more human-like in their capabilities, being endowed not only with the ability to learn but also with artificial emotions, moods and personalities, it will become increasingly less clear that they should be treated as products in the normal sense of the word. And if conventional product liability laws are insufficient to deal with intelligent learning robots, what additional or alternative laws *should* apply? This is very much an open question and in my opinion is likely to remain so for many years to come.

1.4.4   When Robots Are Modified or Misused

Misconduct by a robot, whether accidental or otherwise, can occur when a robot is misused or when it is modified by an owner or user in such a way as to make the product more liable to cause an accident. Clearly any modification made to a robot after its sale might be relevant in how a court determines questions of product defect or the issue of causation. And, modifications to a product could also introduce the possibility of shared responsibility for a robot accident, adding yet another level of complexity to the legal arguments.

1.4.5   Other Considerations of Legal Responsibility

I am sure that by now you will have realized that the issue of product liability in law is far from straightforward, especially in the case of robotic products. And the less straightforward it is, the more the arteries of the courts will become clogged, and the bigger the potential treasure chest will be for the lawyers.

So we must accept that, in cases involving the use or misuse of an autonomous robot, the legal problem of who to blame is a highly complex one, far more so than with a non-thinking product. And the fact that an autonomous robot bears some of the mental characteristics of a human being creates a very different class of legal problems if the robot is faulty.

Believe it or not the complications do not end there. Up to now we have considered only cases where an accident was caused directly by the use or misuse of a robot, but it is also possible for a robot (or an animal) to cause accidents indirectly, for example by distracting drivers or other road users. This type of situation is addressed by German law, under which dog owners have a legal obligation to control the behavior of their dogs on the street. If a dog in Germany irritates or distracts a road user to the point of causing an accident, then the owner of the dog is held to be legally responsible for the accident.

1.4.6   Summary of Legal Considerations

With all of these myriad legal complexities to consider, and all the technological complexities, can it really be reasonable to expect lawmakers to formulate laws that adequately deal with cases of robot accidents? And are judges and juries going to have the necessary background knowledge and understanding to decide justly on questions of liability? In my opinion the answer to both questions is a resounding "No!"

In practice it might be extremely difficult, or even impossible, to fairly apportion the blame for a robot accident and thereby determine which person(s) or organization(s) should pay what proportions of the legally assessed damages. So if there is a robot in every household in some countries, as is planned for South Korea in fewer than ten years from now, consider just how many court cases will occur in relation to robot accidents. The courts will be utterly swamped, not to mention the enormous difficulty and cost involved in determining fault in every case, and especially in those cases where a robot can learn and think. Very nice work for the lawyers, but not much joy for the courts system.

No matter how society chooses to deal with questions of liability for robot accidents, it is inevitable that robot owners will have certain legal obligations to control the actions of their robots to the extent they are able to do so, in much the same way as automobile owners are obliged to control their vehicles. But I believe that laws can only be expected to play a minor role in resolving questions of blame and the awarding of damages in relation to robot accidents. Instead a completely different approach is needed - one that is relatively straightforward and inexpensive to administer and which will not overwhelm the courts.

1.5 ROBOT CARS

Now let us consider robots of a particular type, namely robot cars. The "Grand Challenge" sponsored by DARPA has encouraged rapid developments in the field of driverless vehicles. These are the cars of the future, and they are indeed robots — they are programmed to drive themselves from A to B, recalculating their route as necessary and avoiding obstacles in their path.

Before many years have passed robot cars will be the rule rather than the exception. Human drivers will become mere passengers, avoiding all the hassle and responsibilities related to driving and parking, and commanding their vehicles via speech recognition software. Those who are now drivers will be able, as they ride, to chat to their car's artificial persona, to watch TV, or to work, whatever they wish.

From time to time there will of course be accidents involving robot cars, and those accidents will be covered by motoring insurance in the usual way - every owner of a robot car will be legally compelled to take out adequate insurance cover against risks to themselves, to other drivers, to passengers, to bystanders, and to people's property. This type of insurance is absolutely standard and compulsory for motor vehicles.

The South Korean government goal of having a robot in every household by 2020 puts robots on a similar footing to motor cars as consumer products. In the UK, for example, the average number of cars per household is approximately 1.1, very close to the planned per-household robot population of South Korea. So it makes good sense to consider a scheme for dealing with robot accidents akin to that for dealing with car accidents. The various motor insurance schemes throughout the world fulfill a very important function and by and large do so rather well. And despite there being a huge number of motoring accidents every year, and therefore a huge number of payouts being made by the motor insurance industry, insurance is very big business and mostly a successful business. This, I believe, leads us towards a cure for the cancerous disease called litigation that I have been talking about today.

If accidents involving robot cars can be satisfactorily compensated through compulsory insurance, why not accidents involving just about every conceivable type of robot? Indeed, consider what would happen if motor insurance did not exist, but instead the compensation of car accident victims had to be dealt with by the courts. This would bring about the immediate collapse of the court system in every country throughout the world. For this reason alone it seems to me that compulsory robot insurance is the only sensible way forward.

1.6   INSURANCE

Recognizing the many difficulties that come with the prosecution of robot accident cases, some recent authors have proposed insurance as an alternative remedy. Insurance claims are generally much easier and far less costly for the public to manage than are lawsuits, which is an important point in favour of insurance, rather than litigation, as the most appropriate ameliorator following a robot accident.

The ideal robot insurance policy would provide financial protection against physical damage and bodily injury resulting from any type of accident involving the robot. Amongst the risks covered would be: medical payments and compensation for anyone injured by a robot (including emotional injury), as well as damage to the robot itself if caused by another robot, and damage to other property.

I am in favour of no-fault insurance as an essential element of a robot insurance scheme, in other words one which pays out irrespective of whose fault an accident might be. Such schemes typically lower the cost of insurance premiums by avoiding the need for expensive litigation to determine who caused an accident. In fact a condition of no-fault insurance is that the victim, who is paid out by the insurance company, is not allowed to litigate for damages. How sad is that for the lawyers?

It is the compulsory nature of vehicle insurance that provides the greatest safeguards for compensation to anyone who suffers injury or material loss as a result of a motoring accident. In order to enforce these requirements on the driver of a motor vehicle there are normally heavy legal sanctions against drivers who are not adequately insured or who are not insured at all. In the United Kingdom the Road Traffic Act makes it a criminal offence to drive a motor vehicle on a road or some other public place without adequate insurance cover, and it is also a criminal offence to allow or encourage someone else who is uninsured to drive. The primary purpose of that legislation is to ensure that funds are available from insurance companies to compensate the victims of driving accidents. But if a driver is uninsured there is no readily available source of funds from an insurance company to compensate the victims of that driver's accidents, so in the UK there is a voluntary agreement between the government and the Motor Insurers Bureau, to provide sufficient funds to deal with such situations. This agreement not only provides for cases where a driver is uninsured, but also for accidents incurred by drivers who abscond and are not traceable for the prosecution of an insurance claim.

I support the proposal by Anniina Huttunen and her colleagues in Finland, a proposal also discussed by Ryan Calo at Stanford University, for a compulsory insurance system for robots. But I go even further by advocating an electronically driven monitoring process for law enforcement agencies, whereby they can check if a robot is adequately insured. If the detection method discovers that adequate insurance is not in place the robot will be temporarily disabled by an electronic message transmitted remotely and automatically, by the authorities, to its black box, so that the robot will not function, beyond announcing to its user that its insurance cover is inadequate or has expired. The monitoring system should not only disable uninsured robots, it should also report to the authorities any attempt to bypass the monitoring technology.

My proposal is that, when a robot is sold by its manufacturer, an appropriate insurance premium should be levied for its first year of operation. That premium is included in the robot's selling price and until it is paid the robot will not function. One month or so before the paid insurance premium expires the robot's owner would receive a warning message, from the insurer, that its insurance needed to be renewed. When the renewal payment is received by the insurer an electronic message is sent to the robot's black box, instructing it to permit the continuing operation of the robot. But if the owner fails to renew the insurance by the due date the robot will cease to function, instead announcing "Please renew my insurance before you attempt to use me again."

In considering insurance coverage for robots, Ryan Calo proposes that

---

"The level of insurance should depend on the nature of the robot being insured. Many robots — for instance, small robots used primarily for entertainment — would only need to be insured minimally, if at all. Larger robots with more autonomous functioning — for instance, security robots that patrol a parking lot — would require greater coverage."

Calo suggests that those who use robots for relatively dangerous activities, such as house perimeter security, should probably purchase substantial insurance coverage, whereas those who purchase robots largely for a sense of companionship need take out less coverage, if any. Other factors suggested by Calo that could affect the level of insurance required could include the presence of children or pets in the house or the overall likelihood that the robot will come into contact with strangers.

Huttunen points out that in the early days of a robot insurance system it will be necessary to provide information about the possible risks so that the insurance market will be able to determine the price for the risk, in other words the appropriate insurance premium. Since it will take some time to build usefully large databases of historical robot accident data, it is likely at first that robot insurance schemes will be expensive in order to protect insurers against errors in their risk profiles. Eventually though, sufficient historical accident data *will* be available to create a stable market.

One of the beneficial effects of employing a wealth of historical robot accident data as a guide to setting insurance premiums, is that robot developers and manufacturers will feel the effects on their corporate bottom lines if their products are more accident prone than the average. The robot insurance system I envisage will be self-regulating for the robot industry, as makes and models that are accident-prone will rapidly attract higher insurance premiums, thereby pushing up their retail costs and encouraging consumers to purchase products with better safety records. Similarly, robot owners will have a financial incentive to take care in how they use their robots, since they could suffer higher premiums through the loss of their no-claims bonuses.

In order to enforce mandatory robot insurance requirements on the owner of a robot, there will need to be heavy legal sanctions against owners and/or operators of robots who are not adequately insured or who are not insured at all, but who have somehow managed to bypass the disabling technology in their robot's black box. If a clever computer expert manages to hack into the robot's insurance renewal software and enable an uninsured robot to be used, they will be committing a criminal offence in the same way that car drivers in most jurisdictions are committing an offence when they drive without adequate insurance cover. The robot's black box, while monitoring the status of the owner's insurance policy, could transmit an alert message to the appropriate regulatory authorities, advising that an uninsured robot is in use, or an uninsured modification to an already insured robot, together with the owner's name and contact details. The automatic generation of penalty notices for certain motoring offences is already commonplace, so a similar scheme for uninsured robots seems eminently reasonable, and in the most serious cases a message transmitted to the police to assist them in identifying and tracking down errant robot owners.

In the case of motor insurance, in many jurisdictions there are some types of vehicle that are deemed not to be motor vehicles for the purposes of insurance requirements. In the UK this includes lawnmowers, and some electrically assisted pedal bikes. These exclusions provide a model that can readily be extended to robot insurance. Some toy robots designed for children, for example, will not need to be insured because the risk of serious injury or death being caused by such a toy is very slight, and warning labels can easily be affixed to the toy to cover such possibilities as swallowing removable parts.

One challenge to a system such as I propose is presented by consumers who, after their initial purchase, subsequently buy hardware or software add-ons that affect the risks posed by their robot. In order to cater for these possible dangers a robot's black box could be programmed to detect any such add-ons and send an electronic message to the insurer's "actuary" — a piece of software rather than a human actuary — which would then calculate the appropriate change in premium, and the owner would be informed accordingly. Even more dangerous could be robot enthusiasts who possess the technical knowhow to modify their robots in all sorts of ways, again creating the possibility of an increased risk, by turning an innocuous robot into a dangerous one. In such cases the robot's owner would be compelled by its black box to self-authenticate by filling out an electronic form with details of the modification, in order that an appropriate increase in insurance premium could be calculated. An owner who is found to have given false information regarding their modification would be liable to prosecution in the same way as a motorist who gives false information on the application form for a motor insurance policy.

1.7 SUMMARY AND CONCLUSIONS

The expansion of the robot society will carry with it a massive increase in the number of robots of every conceivable type, and a corresponding increase in the number of robot accidents. Entertainment robots will flourish as consumer products and therefore accidents in the home and places of entertainment will account for a significant proportion of all robot accidents.

Since the mid-1980s almost all of the discussion on how society might react to the robot accidents of the future has been based on the assumption that the remedy should lie with the law courts. There is a growing literature on the apportionment of legal responsibility for robot accidents, and a small but informative body of case law, both of which have convinced me that, if left unchecked, the litigation of robot accident cases will inevitably swamp the courts in all countries where robots are plentiful. One of the negative effects of all this litigation is that the growth of robotics as a research field and as a branch of commerce will be stunted because commercial robot development, manufacture and marketing will become such risky businesses.

Numerous reasons for doubt, confusion and argument are likely to be put forward in robot accident cases, mostly relating to the liability, or lack of it, of one or more of the suspects I have mentioned today, including:

a. The person, store or other business that sold, hired out or leased the robot.

b. The factory that made the robot.

c. Those responsible for testing the robot.

d. The designers of the robot hardware.

e. The designers and programmers of the robot's software.

In assessing the responsibility or lack of it of all these parties and more, the courts would also have to take into account certain factors relating to the type and characteristics of the robot, including:

f. Its level of "intelligence."

g. Had the robot's hardware or software been modified in a manner likely to have contributed to the accident?

h. Did the robot have the capability to learn and/or to modify its own behaviour in a manner likely to have contributed to the accident?

And the legal quagmire thickens as we consider the possible effects on the outcome of a robot accident trial, caused by any or all of the following:

i. A judge who understands little or nothing of robot technology.

j. Jurors who understand little or nothing of robot technology.

k. Expert witnesses who, while having a good general grasp of robot technology, cannot be expected to have sufficient specialist knowledge to determine the exact causes of the behaviour of a robot that led to a particular accident.

In the face of so many potential obstacles to a fair outcome in a court case, and considering how swamped the courts would become with the inevitable multitude of robot accident cases, it is clear to me that litigation cannot be the answer. But the idea of mandatory robot insurance, based on the no-fault insurance model, has some very significant advantages:

a. It will be relatively easy to administer.

b. Technology will monitor and enforce the legal requirement for owners to be adequately insured.

c. Robot owners will find it financially appealing to choose safer makes and models of robot, because of the lower insurance premium add-on.

d. Robot developers and manufacturers will lose sales if their products are relatively unsafe, since they will attract higher insurance premiums

We can see the risks already. Let us plan to insure against them.

1.8 BIBLIOGRAPHY

1. Asaro, P., 2007. Robots and Responsibility from a Legal Perspective. Workshop on Roboethics, within the 2007 IEEE International Conference on Robotics and Automation (ICRA'07), Rome. 2007.

2. Calo, R., 2011. Open Robotics. *Maryland Law Review*, vol. 70, pp. 571-613.

3. Chopra, S. and White, L. 2011. *A Legal Theory for Autonomous Artificial Agents*. University of Michigan Press, Ann Arbor.

4. DiPerna, P., 1984. *Juries on Trial*. Dembner Books, New York.

5. Freitas Jr., R. 1985. The Legal Rights of Robots. *Student Lawyer 13* (January 1985), pp. 54-56.

6. Hallevy, G., 2010. The Criminal Liability of Artificially Intelligence Entities. *Akron Intellectual Property Journal.* No. 2, pp. 171-307.

7. Huttunen, A., Kulovesi, J., Brace, W., Lechner, L., Silvennoinen, K., and Kantola, V., 2010. Liberating Intelligent Machines with Financial Instruments. *Nordic Journal of Commercial Law*, No. 2.

8. Kelley, R., Schaerer, E., Gomez, M. and Nicolescu, M. 2010. Liability in Robotics : An International perspective on Robots as Animals. *Advanced Robotics*, vol. 24, pp. 1861-1871.

9. Lehamn-Wilzig, S., 1981. Frankenstein Unbound: Towards a Legal Definition of Artificial Intelligence. *Futures*, December 1981, pp. 442-457.

10. Schaerer, E., Kelley, R., and Nicolescu, M. *Robots as Animals: A Framework for Liability and Responsibility in Human-Robot Interactions.* 18th IEEE International Symposium on Robot and Human Interactive Communication, Toyama, Japan.

11. Wein, L., 1992. The Responsibility of Intelligent Artifacts: Toward an Automation Jurisprudence. *Harvard Journal of Law & Technology*, vol. 6, Fall 1992, pp. 103-154.

12. Wilks, Y., 1985. *Responsible Computers?* Invited contribution to panel on Computers and Legal Responsibility. Proceedings of 9th International Joint Conference on Artificial Intelligence, Los Angeles, pp. 1279-1280.

13. Ziaja, S., 2011. *Homewrecker 2.0: An Exploration of Liability for Heart Balm Torts Involving AI Humanoid Consorts.* International Conference on Social Robotics, Amsterdam, December 2011.

II

Philosophical Aspect of Cognitive Robotics

CHAPTER 2

Designing Modular AI Robots Inspired by Amerindian Material Culture

Doros Polydorou

University of Hertfordshire, United Kingdom.

Alexis Karkotis

Bornanidea Lab, United Kingdom.

CONTENTS

C OMPARED to Western countries Japan is considered to be at a more advance stage of AI & Robotic development. One of the main reasons this can be attributed to is because Japanese society, mainly due to religious and cultural sentiments, is more robot friendly than Western cultures. The authors, in this chapter, attempt to firstly identify and list some of the cultural characteristics that makes the Japanese more tolerant towards a human-robot integrated society and secondly to identify what other ethnic groups, with similar socio-religious background and beliefs as Japan and with little exposure to western culture can teach us about interacting with what -western society-refers to as soul-less objects. Particularly the authors focus on Indigenous cultures in Latin America, on their material culture, the societal practices that surround the creation of objects and on their artful and loving attention they place in their production. Inspired by Amerindian community models and their material culture, the authors propose/argue for a move towards person-alized home build robots while offering/outlining/sketching an egalitarian social model for robotic communities.

2.1   INTRODUCTION

Cross-cultural comparisons in the modalities of AI Robot development between countries in Asia and the West indicate that religious sentiments and world-views implicitly inform robotic advancements. In the West, and despite the establishment of secular nation states, the millennia old eschatological religious narratives of "resurrected salvation" implicitly direct AI Robotic engineering along a transcendental discourse where, in its most positivist take, a disembodied mind is projected to roam frictionless in an informational cyberscape of a sort (Geraci, 2006). With the MIT Media Lab and firms like Adept Technology excluded (while setting aside the military sector), scientific research in Western countries — especially since the 1985 "shake out"[1] — has been largely employed to enhance the computational speed and the artificial intelligence software aspect of AI technology. Research labs across Europe and the US appear incurious about engineering the actual morphological features of AI Robots — their nuts, circuits and bolts — lagging behind in their development. Hence commercial availability of robotic technology in the West does not differ so much to what existed 20 years ago. A trend, which given recent developments, will most likely reverse in the very near future as corporations like Google and Apple are becoming once again directly involved with the design and commercial production of Robotic technologies to target Western consumers.

For Westerners, the prospect of creating Anthropomorphic AI [Bio]Robots is a discourse tightly interwoven with the Judeo-Christian creation story (where God created man in his own image and breathed life into him) and as with the field of "Transhumanism" it is generally regarded and treated as a taboo. "Taboo" in the sense that it induces an existentialist reflection as to whether scientists, designers and engineers are indeed acting in a manner resembling God and whether that is appropriate. Myths as anthropologist Clifford Geertz (1966) writes are concurrently a "model for" and a "model

---

[1] A year when the Five Major US Giants, General Motors, General Electrics, Westinghouse, IBM and United Technologies either sold out or closed down their robotics operation departments. This was followed by the entry into the US of foreign manufactures who set up distribution and manufacturing facilities, prominent among them being ABB Robotics, Yaskawa, Electric Corporation, Kawasaki Heavy Industries and Fanuc (Nof, 1999)

of" reality[2]. In this sense we could say that the Christian conundrum in this current era of scientific and technological revolution is whether to symbolically equate the development of AI Technology with either (a) eating once more the forbidden fruit of "the knowledge of good and evil" — an event that got humanity expelled from the garden of Eden or (b) with a return to paradise 'after a long socio-evolutionary journey' to partake from the second forbidden tree in Eden, of life; which in modern times and in scientific terms can translate as achieving a prophesied scientific "singularity." Under the specter of possible sacrilege, a network of religious myths, sci-fi fables and frankensteinian urban legends dissect the multiple possible ways via which anthropomorphic cyborgs and semi-immortal AI robots might turn against their makers once they assume 'agency.' Design concepts of actual morphologies and behaviors of future AI Robots within the West mostly appear as pixelated fictional characters across the cultural mediascapes of intersubjective imagination of audiences rather than as a tangible grassroots design movement in multi-disciplinary creative labs (with relatively ample funds at their disposal). Which is usually the case in Asian countries, Japan in particular. Japan who is leading the way amongst post-industrialized nations in Robotic development accounting for 52% of the world's share of operational robots.

With its long tradition of Shintoism and Zen Buddhism forming the substratum of popular religious beliefs (Maraini, 1983: 52), Japan is one of the very few industrialized countries where widespread discussion on animism is observed in ordinary intellectual discourse "inspiring alternative conceptions of nature, science and politics among Japanese intellectuals, artists and activists" (Jensen and Blok, 2013: 97). The Japanese embrace robots as living things - without the existentialist Mind/Body and Nature/Culture "anxiety" observed in the West (see for example Sahlins, 2005), as popular culture does not promote the idea of AI Robots taking over the world. On the contrary Japanese popular culture has long embraced robots as "benevolent, friendly and cute and as entities which wish to be human". Malevolent forms in turn are rarely observed (Robertson, 2007). Japanese scientists, do not reject animistic beliefs prevalent within Japanese culture and are perfectly comfortable with the idea of robots as living entities endowed with a vital force that can be employed to for facilitate wellbeing. What is acknowledged is that as human creations robots can be good or malevolent, as well as very fragile with surveillance not being, yet, so much an issue as mass media and internet blogs do not generally fear or address it as a problem (ibid.)

---

[2]They are "a model of" — reality inasmuch as what is stressed is the manipulation of their symbolic structures so that they reflect, parallel or resonate with the social reality at hand, or as Geertz writes, the "pre-established non-symbolic system". They are "a model for" — in as much as "what is stressed is the manipulation of the non-symbolic systems in terms of the relationship expressed in the symbolic" so that the myth (as a symbolic system which stores meaning and information (1957:422) is a model under whose guidance physical relationships are organized: it is a model for reality" (1966:6).

Shintoism, which can be regarded as a specific form of animism (Clammer, 2001), is a multifaceted and complex set of cosmological beliefs that interweaves notions of purity and impurity, sincerity and evasiveness, and fecundity that synergistically advocate communal regeneration and harmony with nature. As religious scholar Robert Geraci (2006) explains, "[The] fundamental unit of the sacred in Shinto is the *kami*. *Kami* can be the entities of mythology, the objects of shrine worship, aspects and objects of the natural world, and even human beings." An animist notion which has been extended organically to include 'Social AI Machines' as Shinto (akin to Buddhism) regards nature as sacred and all entities as creative nods expressing this sacredness. "All that which stands out, which inspires awe, is *kami*" Geraci writes, adding that "In the modern world, even if industrialization has polluted it, the natural world remains worthy of reverence and the human being remains a part of that world. Although modern Japanese may not label themselves as "Shintoist" the religion appears in their closeness to nature' (ibid: 7).

Buddhism in turn, differing from Shinto, embeds humanity with the most elevated status in its cosmology since it is only through the human form and via the human experience that state of nirvana is achieved. Though all beings as well inanimate entities possess Buddha - nature, it is humans who are endowed with the potential to achieve Nirvana; that is to consciously realize this Buddha - nature and escape while living from the cycle of transmigration. In what is an inherently a non-dual and yet extremely complex religious system (with an array of symbolic ritual practices) humanity is endowed with transcendental purpose that concurrently advocates compassion to all beings. As opposed to Abrahamic faiths, both Shintoism and Buddhism afford sanctity to robots, with the clearest example being the ritual rites performed by Shinto and Buddhist priest to industrial robots during the early days of their introduction in the 1980s.

Social robots being introduced today, anthropomorphic or multimorphic, are not outcasted nor alienated as foreign "others" but accepted as entities occupying a position within a non-hierarchical cosmology. Gods, demigods, spirits, humans, animals and trees are considered to be equal and to co-exist symmetrically in a universe without a definite beginning and a prescribed final. It is within this socio-religious and highly symbolic net, that social robots emerged and find their place. In contrary to Japan, if we study the history of Western culture and Western philosophical analysis of the mind, we can see a dualist (body/mind) approach beginning with the ancient Greeks and reaching its peak with Descartes. The appearance in turn of manmade Automaton machines is not a new invention/concept within the West, as there evidences of automatons as early as the 4th century BC. Nor is new the suspicion that Westerners hold towards their development. Hepheastus created Talos to defend Crete, only to be destroyed by a lightning bolt cast by Media at his single vein of lead. Doctor of the Chuch Albertus Magnus created a man of brass in the 13th century rumored to have such a capacity to reply intelligently to complicated questions, something that deeply disturbed the theologian Thomas Aquinas who beat the bronze man to pieces with a hammer. Pygmalion, the great Cypriot sculptor, carved a woman out of ivory and with the help of Aphrodite changed her into a real woman.

2.2   EVOLUTION OF TECHNOLOGY AND IDEAS

In Japan the introduction of AI social robots in almost all areas in life is desired and anticipated, both on the popular and the governmental level. One of the main targets of Innovation 25, Prime Minister Abe's vision of redesigning Japanese society by 2025, is reversing the declining birth rates on one hand and to accommodate the needs of a rapidly aging population on the other, "[emphasizing] the central role that household robots will play in stabilizing core institutions, like the family" (Robertson, 2007). With a stagnated birthrate of 1.3 child per woman, almost at par with death-rate figures, and 21% of the population over 65 years of age, a statistic which will double in 20 years if the demographic trend continues, the Japanese government has identified "science and technology" as the platforms to reverse demographics and re-boot the economy. Robots are projected to do the share of the domestic tasks, taking care of the children and service as babysitters freeing Japanese women to more eagerly pursue marriage increasing thus the desire to have more children, (re)stabilizing in the meantime the traditional extend family. The majority of Japanese elders in turn seem to prefer "homemade" humanoid Japanese robots as caregivers rather than immigrant foreign workers, because of culture intimacies.

One example of incorporation of robots into Japanese daily life is robot teacher Saya (Hashimoto, 2011) which can express six basic emotions including scolding students when they misbehave. Jordi Vallverdu in his paper 2011 paper "The Eastern Construction of the Artificial Mind" states that we must take into account that Eastern societies have greater IT and robotics implementation and social integration that Western ones and gives as an example the "Robot Ethnics Charter," a legal regulatory system created for robots in Korea in 2007. Jordi believes there are two main reasons for this:

"To prepare a future establishment of ethical guidelines for the partnership between robots and people and to establish Korea as a test bed country for robots," adding that "There is another reason, a deeper one: they love robots, they do not look upon them as competitors and they wish to have a robot in every stage and position in their lives." Bartneck has shown that people's perception of robots relates to what is perpetuated by media and entertainment industry (Bartneck, 2007). Popular culture in Japan and especially anime and manga often portray robots as cute or Kawaii (a notion which reflects popular sentiments of beauty), and this captured the imagination of children from a very young age. In her PhD thesis Annihilating Difference? Robots and Building Design at MIT robot anthropologist Kathleen Richardson demonstrated the Japanese employ robots in multiple social spheres for every aspect of life, from child care, to communication, therapy, cleaning, heavy industries, security and entertainment (Vallrerdu, 2011).

In the contrary, Robert Geraci in his paper: "Spiritual Robots: Religion and Our Scientific View of the Natural World," argues that many researchers working in the US are directly or indirectly influenced by their culture. Their perspectives, Geraci writes, "stems from a Christian tradition emphasizing the need to transcend human life" (Geraci, 2006 p.2). AI researchers in the West are heavily concentrating on cyberspace and their desire to build a heavenly kingdom in virtual reality. To emphasize his point Geraci critiques the argument of architect and software engineer Michael Benedikt who envisions healthy social structure exclusively in Cyberspace, going as far as saying that "the vision of the Heavenly City is a....religious vision of Cyberspace" (Benedikt,1991 p.16). "In order to attain the heavenly city human beings must give up their bodies. Virtual reality holds out the idea of dissembodied paradise" Geraci writes, adding that virtual reality authors in the 1980's looked forward to the eventual (virtual) release from the constraints of the human body. (p.3). Popular culture reflects these ideas with movies such as *The Matrix* trilogy, *Tron* and *Johnny Mnemonic*. Thinking about the fact that the term robot was coined in a Chezh play performed in 1921, where robots finally rise up and kill their human creators, it is perhaps no surprise that American robot researchers generally concentrate on robots with military applications!

2.3   THE AMERINDIAN WORLD AND THE TECHNO–ANIMISTIC PARADIGM SHIFT

Japan is one of the few industrialized countries following an animistic way of life. There are however other cultures whose people adhere to similar beliefs. The authors, especially fascinated with the vast differences between eastern and western understanding of robotic cultures, decided to investigate and refer to other smaller ethnic groups with alternative animistic believes with the aim that their alternative views can shine new light into the theory of technological embodiment. In particular we analyze how animist indigenous communities in Latin America, with minimal exposure to Western cultures, create communities based on trust and egalitarian principles while emphasizing "individuality" as a central societal principle.

Through this analysis we attempt to delineate new approaches in the development of AI in post-industrial societies so that emerging robotic technologies can have a positive effect in preserving traditions and facilitate feelings of communalism and mutualism on both the household and community level. Furthermore, the authors will offer some insights and suggestions deduced from the Amerindian way of life and world view in the form of possible suggestions that could be taken into consideration in the future development of emerging technologies. In animistic societies like the ones we present below, individuals do not simply employ another person to manufacture objects embedded with societal significance and symbolic meaning. What the authors propose in the present article is to make robot parts re-usable so that they retain their uniqueness when updated models are produced. Before we continue, it is important to identify certain characteristics of how Amerindian societies view the world. These observations are mostly made from the personal experience of one of the authors who conducted ethnographic fieldwork among the Ngobe indigenous people of Panama for a period of 3 years (Karkotis 2012). The first question that we would ask is: Why do Amerindian societies offer a good perspective on technological embodiment and how does animism as a worldview offers a perspectival platform to interact with emerging technologies in the future.

As it appears, in the near future, cybernetic tools with capabilities of personalising and augmenting one's reality through bio-digital holographic forms will further complicated the gordian webs of significance which humans weave and find themselves suspended in. Cyber gadgets will likely enable humans to personalize how they perceive the world, how they interact with it and within it and how they appear (and disappear) to others. An individual embedded with cyber-lenses may choose to see the sky as red and pixelated, the sun as green and the moon in a blue. Digital tropical birds will fly across a bird-less sky in Paris and lovers will be falling in love under a virtual starry night in broad day light. Biology teachers of the near future will walk with their students right into Simulacrum holographic theatres of different (Pro)Eukayotic cells, walk around their nucleus, feel the environments, sense their membrane, toy with their ribosomes, swim in their cytoplasm, observe their growth and metabolism, monitor their division and replication.

All this may sound like science fiction. But if it does is probably because the concept itself is not new. As a matter of fact it, sounds very much like animism and paganism: future age, new age, old age, ancient age. Touching an apple and feeling its energy, reading messages in cloud formations and perceiving spirits are practices observed all around the word. Emerging technologies will not replace New Age crystal healing or Voodoo or other syncretic religions in the Americas, as well as, JuJu practices in West Africa, nor will they replace complex belief systems in spirits by forest dwelling societal groups in Latin America. As anthropologist David Graeber writes, "The West might have introduced some new possibilities, but it hasn't canceled any of the old ones out" (2004). Yet a lone Lakota warrior riding his horse on Venus in search for potential ritual sites and a Zen Master holding satsang in a garden on a space station outside our solar system is not exactly how the West, at least, envisions the future. Anthropologist Joanna Overing (1995), emeritus professor and head of the St. Andrews department of Social Anthropology for many years, writes in regards to Amerindian cosmology, itself being very similar with animist cosmologies found in Africa and Asia,

In Native Amazonian ontology there is no "nature", no inanimate or value free universe over which humans can dominate. The notion of "nature"

> belongs to the Western paradigm of power, and not to the Amazonian one, where other worlds are always filled with agency with whom it is necessary for humans to deal. In contrast, there developed in Western understanding the idea that there exists a "natural" order of submission between humans and other beings or elements of the universe [and thus as] disengaged humans, they then had the right to conquer the world of "nature" toward the end of producing wealth for themselves. (:199)"

In an animist cosmology the entire landscape is already embedded with an animist schematic layer, or interface. Even a piece of rock can be inspirited and personified in what anthropologist Viveiros de Castro (1998) terms Amerindian "multinaturalism and perspectivism." The common conception amongst Amerindian people, "whereby the world is inhabited by different sorts of subjects or persons, human and non-human, which apprehend reality from distinct points of view" (:469). Where "in sum, animals are people, or see themselves as person [and their manifest form] a mere envelope (a 'clothing') which conceals an internal human form, usually is only visible to the eyes of the particulars species or to certain trans-specific beings such as shaman" (1998:470). Given the propensity of indigenous people worldwide and through the centuries to appropriate western tools that are considered either useful or simply desirable, it is safe to speculate that in the future, an AI robot that safeguards indigenous legal rights, participates in everyday community activities and is a safe-keeper of oral tradition will most likely be accepted as a member of the community; a non human "other" with agency, who, like all non human entities retains the potential of exhibiting personhood. The form the robot would assume would ideally be, in this speculative future ethnographic scenario of ours, for the indigenous people themselves to design and make.

2.4   ANIMISM: MATERIAL AND METAMOPRHOSIS

In his book Perceptions of the Environment (2000) Tim Ingold fuses together ecology and phenomenology to argue that societal culture is in many respects the weaving together of material objects in a process of "emerging involvement" within the "lifeworld" and that life itself is woven together by a web of movements. Weaving a bag for example is multi-layered process with deep significance: it is an act of mobility, one of education, a practice of enskillment, a way of knowing, a process of storytelling and a ritual. Amongst the Ngobe indigenous people of Panama, the act of weaving a netted bag is exactly such a multi-dimensional process and is embedded with a mythology which guides the weaver's fingers during its making. Households harvest the leaves of the *pita* plant a subterranean bromeliad (Aechmea magdalenae), and after a long process extract its fibers, known as kiga in the local language. The thread from pita (used throughout Central and South America) is also used to make ropes, nets, fans, hammocks, and even strings for musical instruments but it is the weaved bag called *chácara* which carries both life: their babies as well as the source livelihood: wood for fire and food (tubers, plantains, rice) and as such it is enveloped with a ritualistic aura.

"The process" the Ngobe women emphasize "must be done with a clean mind or the result of one's weaving will lead one to face the (same tragic) end that Mesi Kiga faced" the mythological archetypical figure of "ancient days"that taught the Ngobe the art weaving and who stands today as it is patron. In effect, any anxiety, or stress or egoism during its manufacture may bring dire results. Anxiety attracts 'anxiety' spirits (chela) to inhabit the threads that make the bag, which thereafter may transfer and pollute the food consumed which can henceforth bring decease to the household. This notion is also extended to include most activities in life. Eating food while stressed for example is thought to bring illness and even death. So much so that if a person was a food victim, long discussions will follow, disseminating whether it was the food itself that was "bad" or the consumers' "thoughts."

The important point to be made is that the woman has to be in high spirits to either teach the craft/chaácara or make one. Such practice such require tremendous skills and constitutes one of the most important technical practices in the "everyday life" of people living in stateless societies, part of a set of practices embedded with almost mystical and metaphysical significance (Overing, 2003). Therefore it is perfectly understandable that certain material objects, especially the ones which are handmade or handcrafted are very important and invaluable to members of Amerindian societies. This directly contradicts western societies where 95% of the objects used are mass produced and fall under a manufacturers planned obsolesce. Objects in Amerindian societies which are produced by the individuals themselves hold so much value that when they wear down they are often turned into a different object which is equally loved and respected as it doesn't loose the meaning it had when it was originally made. Western countries have a tendency and in fact a celebrated trend, to change or "upgrade" our gadgets every year. That involves actually throwing the most beloved and used objects, a cellphone, in the garbage only for the prospect of owning the new iteration which was just released, transferring the data from the old to the new. This begs the question; will the same trend be followed when Social robots became integrated into our society? Will children throw in the garbage the robot that held their hand as they tried to walk, the robot that helped them do their homework and took them to school every day? Or most importantly, do we really want our children to develop an intimate relationship with what is destined to remain a product with a lifespan of 2 years? The first suggestion that the authors would like to proposition is re-usable parts, or put simply the idea of modular interchangeable parts.

Furthermore, apart from the obvious advantages stated in the previous paragraphs (a deeply rooted need to develop feelings for all members of the family that helps us into personhood) having modular reconfigurable systems made by interconnecting multiple similar units can offer a plethora of advantages. Zykov et al. demonstrated a large space of possible robots capable of self-reproduction (2007). Pancheco, in the project Fable a modular robotic playware platform, has developed a set of robotic parts that would allow nonexpert users to develop robots ranging from advance robotic toys to robotic solutions to problems encountered in everyday life (Pacheco, 2013). Yim, described a modular system as " robot made up of a chain of simple hinge joints could shape itself into a loop and move by rolling like a self-propelled tank tread; then break open the loop to form a serpentine configuration and slither under or over obstacles; and then rearrange its modules to "morph" into a multilegged spider, able to stride over rocks and bumpy terrain" (Yim, 2007). Finally, Hamlin, in his work Tetrobot, demonstrated an actuated robotic structure which may be reassembled into many different configurations while still being controlled by the same hardware and software architecture (Hamlin, 1997)[3].

Returning to Amerindian material values, we can identify that one of the main reasons that Ameridian societies value their belongings is because, firstly they are their makers and secondly because they spent ample time to enhance their objects using art and craft techniques to bring to light the myth that envelope(s) their production. As many anthropologist point out, the very act of their making is what promotes and sustains their egalitarian societies. Western and Japanese societies, on the other hand, do not afford individuals the luxury to spent time on personalizing the hardware of their own objects, nor is that feat promoted. It is important here to note that we are not referring about software tasks such as changing the background of a mobile phone–something which can be replicated in minimal time on a new phone, but rather, hardware tasks. Even though they can be suited to the skills of the individual, they would require time, effort and motor activity. Engaging individuals with the actual manufacture or assembly of the hardware of their AI Robot will create more intimate relationship both within the family members and with the actual robot itself.

A very important argument made by Joanna Overing, throughout her work, is that while Amazonian people overtly dismiss the idea of a social rule, they hold at the same time in high regard matters of personal autonomy and strongly value sociality, their traditions, and "the mutuality of ties of community life" (1989, 2000, 2003, 2006). What should be underlined is that emphasis on personal autonomy should not be confused with egoism but as a way of acting independently a community of similars based on the principle of trust and respect for each member (cf. Santos Granero 1991:254 and De Carteau 1984:6). As Overing explains, in order to understand their "treasured peace as well as their egalitarian and informal ways" one needs to appreciate

---

[3]For an analytical description of a number of different projects involving modular systems, please see: M. Yim, W-M Shen, B. Salemi, D. Rus, M. Moll, H Lipson, and E. Klavins. Modular self-reconfigurable robot systems: Challenges andopportunities for the future. IEEE Robotics & Automation Magazine,14(1):43-52, March 2007.

the emphasis that indigenous people place upon everyday creativity, where most artistic production belongs to. Overing writes in regards to the Piaroa people of Peru, with whom she has been conducting ethnographic fieldwork for almost four decades,

---

"'The beauty and tidiness of the implements they make for daily use, the attention they pay to form and design, are striking. These beautiful arti-facts and tools serve to exemplify their work of beautifying as many as-pects of everyday existence as possible. Because their major concerns relate directly to the artful skills of daily life, the Piaroa endow activities that we might see as merely humdrum (preparing a meal, weeding a garden, making a basket, feeding a baby) with significance far beyond any that we might consider. The artful everyday of the Piaroa is linked to a principle of trust, for it is only through the creation of trust that the artful everyday of this egalitarian people can be constructed. (2003:295)"

"Such peoples" she writes "can offend anthropological sensibilities on a number of fronts" because, Western academics are then forced to an-swer, "how can people link social mutuality with an insistent individualism? (2003:298). Part of the problem why this question boggles the "Western 'ur-bane' academic" is because, according to Overing, he or she views everyday matters as tedious: cleaning of the dishes, feeding of the children, weaving a bag, cleaning of the body, cleaning of the house, making a canoe. And, what tends to occur amongst Amerindians living in the jungle, is that matters of everyday concern, such as feeding the children, attending the needs of the house and artistic production, are highly valued and respectfully attended to. They are often embedded within the greater Amerindian cosmological perspective of the world. In post-industrial societies these matters are con-sidered disdainful and should be attended and accomplished as quickly as possible. Overing concludes: "If they can be attended by another person it is even better".

The aporia raised is the following: What would be the societal effects if members of a western society or industrialized society treat a sociable robot as a quasi-embodied entity which at least in its early stages in a family integration will require a lot of emotional involvement and social acceptance - the same way we treat any other electrical appliance in the household? If the task of maintaining or "programming" — for lack of a better word — the robot becomes a mundane task, wont that act as counterproductive in the general acceptance of robot integration as part of our community?

The authors would like to recommend an alternative proposal, which is the customization of robots through arts and crafts. In this scenario, akin to Amerindian families, the Western household would put the effort to become part of the creation process of the robot, even if its just on the decorative or expressionist level, which would develop a much stronger bond both between the family members and with the Robot. In this sense the design of the robot would become an evolving process, its actual form changing in accord to family innovation. This can be even taken a step forward, with bio-customization from their individual users. This can take the form of bio fuel (e.g. waste material from the human body) leading to a deeper personal connection with the robot and a much more intimate relationship.

2.5 EGALITARIAN SOCIETIES

As robots populate countries and their AI capabilities increases to the level of independent thought and existence, it is important to ponder what kind of sub-societies robots will create amongst themselves. What kind of social values and what kind of character traits will they need to display in order to be self-sufficient? Would they resemble a western society consumerism structure? Or is an Amerindian structure better suited for them? The Yanesha people of Peru for example, with whom the anthropologist Ferdinand Santos - Granero conducted extensive fieldwork with, identifies that the ideals of love, friendliness, trust and generosity are social virtues that are taught to everyone from early age and need to be re-enforced on a daily basis. And those individuals who deviate from these ideals are isolated and ignored rather than confronted. As such what is reinforced to prevent isolation from the whole is 'controlled sociability' which sees households making regular visits to one another, which in turn presents the need for autonomy and distancing from the whole, without being isolated from it (Santos-Granero 2007). This distancing from the public sphere without being isolated is presented for example in the houses of the Yanesha which have no walls, so that everyone can see what goes on inside - a pattern observed amongst indigenous people throughout Latin America.

For a household within the typical Amerindian village to oblige with its responsibilities it must thrive to be as autonomous as possible (while retaining fully active its kinship membership). And for this to occur a man or a woman have to be polymathians. What one observes in egalitarian non-hierarchical communities of "similars" are polymathian individuals who know similar traits and area always interested in learning more or expanding the ones the already know; a sort of Amerindian polymathianism. "Similar" in turn does not mean "the same" as much as it means analogous and equivalent. "A social relation" writes Radcliffe Brown (1935:396) amongst South African tribal societies" does not result from a similarity of interests, but rests either on the mutual interest of persons in one another, or on one or more common interests, or on a combination of both of these." Personal responsibility and personal autonomy are notions the Ngobe highly value alongside harmony, friendship, and sister-brotherhood. Anthropologist Philip Young, who conducted decades of fieldwork amongst the Ngobe writes in regards to the matter of responsibility,

"The expectations of mutual aid generally require verifying that another has a desired item, such as a tool, or verbalizing openly the kind of help one needs, like the repair of a boat or the loan of money. The nuances of generalized reciprocity are more subtle than those of balanced reciprocity. Since kinsmen who live in close proximity are involved in everyday kinds of assistance, people expect to receive without having to ask incessantly. The expectations one holds about close kinsmen easily become frazzled and readily provide the grist for night time dreams. Kinsmen may delay too long before providing a favor, compensate with much less than was expected, or circulate a desired item (or give too much) to someone other than oneself or the close kinsman one had anticipated would receive assistance. (1991:187)"

Thus, in the actual social reality of the village, this link between social mutuality and "insistent individualism" is established itself in the multi-practices that each member of the household engages in. It is the multi-micro-quotidian practices which defines household praxis. A household in a typical Amerindian village, retains a sense of autonomy from its kin-group, which may be two - to ten - hours walk away, and from the rest of the household in the village via these micro-quotidian practices. At the same it has the social obligation to retain (or rather constantly regenerate) this autonomy, both with respect in the rest of the village and its kin-group, so that it does not become dependent on others the members of the household practice an array of activities. Thus having autonomy entails responsibility, knowledge and practice, as well as the being able to reciprocate [4].

The wife, for example, apart from cooking, cleaning, educating and nurturing her children, tells myths, farms, helps her husband with the building of their house, helps with the construction of the family canoe, goes to the forest to collect herbs for medicine and gathers plants to weave the chacara bags. She fishes, she gives birth in a hut with her family around her, she can carry fifty pounds of weight barefoot in the jungle. Thus her role is not one-dimensional. She is a mother, a farmer, a botanist and an artist all at once. The same applies to the husband. The children in turn help their parents at almost all chores and learn the Ngobe culture via these practices. At the same time they receive western education as well. With all its members being polymathic and multi- talented the household tries to be as self-sufficient as possible. "Primitive society constantly develops a strategy destined to reduce

---

[4]Gow (2000) describes the Piro notion of "helplessness" in terms of they acknowledging their inability change the "given structure of the cosmos," which induces them into seeing the same "helplessness" in to others, which in turns promotes mutual sympathies, mutual assistance and the emphasis on "living well." "So, the something that does not happen" writes Gow "is helplessness, suffering. The flatness of everyday life turns out to be fully intentional, it is an achievement" (: 76).

the need for exchange as much as possible" writes Clastres (1977), and he is absolutely right.

Amongst the Ngobe, who like many Amerindian groups, they practice subsistence economy, there are (irregular) food surpluses within the village that the household does not and cannot consume and are distributed to family and friends who request it. At some time or another, an agricultural field will either yield a surplus, yield nothing, or yield the normal and sufficient produce. Sometimes the opposite occurs, as was the case with the 2010-2011 floods, that destroyed most of the finca crops in the entire Krikamola region. I had arrived for fieldwork shortly after, and the dismay from the luck of food was explicit in the people's faces. Importantly though during this period collaboration amongst household and kin groups was enhanced. Therefore inter-household food distribution occurs throughout the year. For example if a household has a bad harvest one year, or if its field is simply too small to constantly produce food, its members will request to accompany another household when its members go to the field, help with chores and gather food from it for themselves. They are implicitly performed social etiquettes which occur between hamlets and between households which have no name by which to name them. And this is what makes a gift exchange important and this is what makes giving away food also so important because to be able to "give" implies that one is self-sufficient and able to do so in the first place. This effort toward self-sufficiency — who rests on the principles of 'care' and 'trust' between the members of the family — is the cornerstone for household autonomy and social stability and harmony to be maintained. Joanna Overing (2003) writes regarding the Piaroa people "The stress the Piaroa put upon the everyday and its activities is not trivial, but rather the product of a powerful and highly egalitarian social philosophy".

2.6 FORMATION OF A ROBOTIC COMMUNITY

Samani et all, in their paper Cultural Robotics: The Culture of Robotics and Robotics in culture attempt to define a robot community culture which is "programmed by human, is human-like, and human comprehensible" (Samani, 2013). They propose that such a culture would refer to the creation of values, customs, attitudes, artifacts, and other cultural dimensions among a robot community. The prerequisites for robots to evolve a cultureâĂŞ always based on the human definition — would be an independent, critical and self-reflective mind that develops in a way that leads to consciousness and, ideally, self-awareness of the robot. They continue by identifying two characteristics, which in their opinion serve as the fundamentals aspects in any human culture: (a) repetition of social and individual learning and (b) a Darwinian-inspired theory of selective reproduction. For a family robot to be custom-made by the family itself, would give it a sense of genealogy that is traced back to the family members, which gave it form.

As we discussed in an earlier section, there are currently robots employed in all kind of different industries, including social, manufacturing, entertainment and military. In the beginning, as robots are currently made and optimized to perform specific repetitive but fundamental tasks, it will come as no surprise that they will use the exact same skills in order to contribute to their own community. As they will not possess the four primary human emotions, anger, fear, sadness and happiness (Turner, 2011) they will not have to deal with a surge of destructive thoughts and situations which they could lead to isolation. Turner identifies that in most human societies; the majority of the primary emotions is negative and generally works against forming social ties and groups. People who are angry invite counter anger; those who are fearful are usually avoided. The same is true for most intelligent mammals but unlike humans they have built-in programs to form packs, prides, pods, troops and herds therefore not needing to rely so much on emotions to create and sustain social structures (Ibid). Specialist robots will work together, forming groups and they will quickly build a living space and their infrastructure based on the most functional and efficient use of the space. They will create their own power generators which will be completely independent from the world's natural resources and using their problem solving abilities (in the beginning probably by using trial and error and adapted learning) they will quickly optimize their living environment to fit their needs.

As the need for social robotics increases and cultural acceptance is established, we will see the rise of the "polymathian" and "multi-morphic" robot, as families would prefer to have one robot suitable for more than one task. As these robots become part of the core family group, there will be a need for them to demonstrate feelings of affection in order to become more easily integrated and accepted as part of everyday life. These feelings will also be reflected in both a positive and a negative way in their particular cultural settings denoting the emergence of robotic individuality and creativity.

In order to reflect their newly creative nature, robots will either "want" or be implored to create an identity. Adhering with the second proposition from this chapter, each robot would ideally be customized and personalized by the family that is part of, an perpetually evolving change of form and interactivity. The robots in turn, by learning and adapting, both from their "personal experience" with the human family as well as by each other (in conjunction with the networked information which will be available from the net), they would begin to customize their own living space, valuing differentiation, alterity and simulation.

This in turn, along with other character traits which will spawn from their deviated and individual "thinking" process, may in fact appear to be counterproductive to the social structures of their community, inducing the emergence more chaotic structure. At this point, we can refer back to the Amerindian societies and to the concept of "controlled sociability". What the robots might have to, as a form of counter power against a chaotic environment, is to distance themselves from the rest of the group, in order to give way to their individuality and autonomy, without however becoming isolated from the group itself. As robots are now polymathians, and their learning and mastery of skills will probably be dependent on the kind of work they are performing with the family they live with or around the industry they work at. Just like in Amerindian societies they can offer back these skills , from weaving to care-taking, to the human community for the mutual good. As robots will primarily be made of reusable parts they can open their own robot clinics where they can offer part exchange services. Robots which are skilled in constructions will be able to help with the building and repairing of hardware infrastructure. Robots skilled in telecommunication could serve as network hubs to the outside world and robots skilled in industrial design they could work on manufacturing and building of spare parts, all under an egalitarian community based on the exchange of services and the mutual good. This future vision design of robots cultures would reflect and promote a rhizomatic network of interrelationships, based on mutualism and communalism, with inherent ecological and environmental sensitivities.

2.7 CONCLUSION

As the nature of this topic is by default highly speculative, this book chapter aims to provoke further discussion and offer some fresh perspectives approached by an anthropological perspective. The authors began by examining the current state that led to the evolution of robotics, concentrating more on philosophical thinking behind the western and the eastern cultures. From the background research it was evident that the west and the east have very different thought processes. The western cultures influenced by the body/soul divide and theological and mythological ideologies which demonize the non-human seem much more reluctant to accept robotics as part of their families. The eastern cultures however, with their Shinto and Buddhist traditions are much more open to the idea. The authors, wanting to identify new knowledge and explore what other ideas can be learned from other communities with different religious and beliefs; they offer an investigation into the daily lives of Amerindian societies, focusing on their animist beliefs and their material culture. These communities have limited exposure to western and eastern population which minimizes as a result external influences in their way of thinking. Due to their animistic views, the members of these communities already view the entire ecosystem, its entities and its objects (organic and inorganic) as potentially possessing an interface or a personhood. Energies (as a form of data) and personhood (akin to an avatar) are embedded in the very definition of each object and entity they touch and interfere with.

By investigating how they live their life, the authors have reached to specific examples which they can be directly applied to our thinking process of non-human technological embodiments. The first proposal, spawning from the importance these communities place in the making and crafting of the objects they own and use in their everyday life and subsequently the personal value they place upon these objects, is the idea of creating re-usable parts for our electronic equipment. As we are attempting to create Social robots which they will become part of our everyday life, we need to start giving value not only towards the function of a system but also to making of the system itself. It is understandable that throughout its life span a body part may have to be replaced as new and better technology will come along. However, our design should be concentrating much less on technological obsolesce and the dangerous comfort of merely disposing the old part in order to purchase a newer model, but rather on how we can use most of the old equipment as part of our new system. After all the point in social robotics and robotic acceptance is to create robots which we can trust, invite to participate in our lives and perhaps develop some feelings of intimacy for them as opposed to throwing them in the garbage and get a new one every two years. Secondly, building on the idea that Amerindian societies spent a lot of time personalizing and enchasing their objects by using arts and crafts with symbolic meaning embedding in the pre- and post- production, the authors would like to propose that spending time in order to personalize and create a unique identity for each AI robot should become an integral part of the developmental process. Moving away from the idea of the mass produced, robots should become personalized, even custom made to the individual family if that's a possibility. Finally, the authors offer a vision of how robots can create their own community, based on egalitarian principles and the exchange of services.

2.8 BIBLIOGRAPHY

Bartneck, C., Kanda, T., Mubin, O., & Mahmud, A. (2007). "The Perception of Animacy and Intelligence Based on a Robot's Embodiment". Proceedings of the Humanoids 2007, Pittsburgh pp. 300 - 305. | DOI: 10.1109/ICHR.2007.4813884

Benedikt Michael, "Introduction," *Cyberspace: First Steps*, ed. Benedikt (Cambridge, MA: M.I.T., 1991): 1-26, 16.

A. R. Radcliffe-Brown (1935). "On the concept of function in social science", *American Anthropologist New Series*, Vol. 37, No. 3, Part 1. pp. 394-402. Wiley.

(1974). *Society Against the State* translated by Robert Hurley Mole Editions, Basil Blackwell. Oxford.

Clammer, J. (2001). 'Shinto dreams: difference and the alternative politics of nature'. *Japan and Its Others: Globalization, Difference and the Critique of Modernity*. Rosanna, Vic.: Trans Pacific Press, pp. 217-243.

De Carteau Michel (1984). *The Practice of Everyday Life*. Translation by Stephen Randal.

G. Hamlin and A. Sanderson (1997). "Tetrobot: A modular approach to parallel robotics", IEEE Robot. Autom. Mag., vol. 4, no. 1, pp.42 -50

Geertz Clifford (1966). "Religion as a Cultural System," in M. Banton (ed.), *Anthropological Approaches to the Study of Religion*. New York: Praeger, pp. 1-46.

(1957). Ethos, world-view and the analysis of sacred symbols. *The Antioch Review*, vol. 17 no. 4 (1957), pp. 421-437

Geraci, R.M. (2006). "Spiritual robots: religion and our scientific view of the natural world", *Theology and Science* 4(3): 229-246.

Geraci, Robert M. (2006). "Spiritual robots: Religion and our scientific view of the natural world." *Theology and Science* 4.3: 229-246.

Gow, Peter (2000) Helpless, the affective preconditions of Piro Social Life. *The Anthropology of Love and Anger: The aesthetics of conviviality in native South America*. J. Overing and A. Passes eds. London: Routledge.

Graeber David (2004). Fragments of an anarchist anthropology. Chicago: Prickly Paradigm Press (distributed by University of Chicago Press).

Ingold, Tim. (2000). The perception of the environment: essays on livelihood, dwelling and skill. London: Routledge.

Jensen and Blok (2013). Techno-animism in Japan: 'Shinto Cosmograms, Actor-network Theory and the Enabling Powers of Non-Human Agencies' Theory Culture Society 30:84.

Karkotis, Alexis (2012). Now we live together: Community formation amongst the Ngobe and how "Majority Rule is still an Alien Concept Amongst Most Guaymi" PhD Thesis. Submitted to the University of Bristol.

M. Yim, W-M Shen, B. Salemi, D. Rus, M. Moll, H Lipson, and E. Klavins (2007). Modular self-reconfigurable robot systems: Challenges and opportunities for the future. IEEE Robotics & Automation Magazine, 14(1):43-52.

Maraini, F. (1983). 'Japan, the essential modernizer'. In: Henny S and Lehmann J-P (eds) Themes and Theories in Modern Japanese History. London: Athlone Press, pp. 27-63.

Moises Pacheco, Mikael Moghadam, A. Magnusson, B. Silverman, Henrik Lund Hautop, Christensen David Johan (2013). Fable: Design of a modular robotic playware platform. ICRA 2013: 544-550

Overing Joanna (2006). The Backlash to Decolonizing Intellectuality. Anthropology and Humanism, 31: 11-40.

(2000). "The efficacy of laughter: the ludic side of magic within Amazonian sociality." In The Anthropology of Love and Anger: the aesthetics of conviviality in native South America. Overing J. and Passes A. eds. London: Routledge.

(2003). "In Praise of the Everyday: Trust and the Art of Social Living in an Amazonian Community". *Ethnos*, Vol, 68:3, Sept, (pp. 293-316).

(1996). "The conquistadors of the jungle: images of the Spanish soldier in Piaroa cosmology." University of Indiana, 14:179-200. Berlin.

(1989). "The aesthetics of production: the sense of community among the Cubeo and Piaroa" *Dialectical Anthropology* 14: 159-175.

Ph.D. Thesis. MIT University: USA.

Robertson Jennifer (2007). ROBO SAPIENS JAPANICUS: Humanoid Robots and the Posthuman Family. Critical Asian Studies 39:3 pp.269-398.

Samani Hooman, Elham Saadatian, Natalie Pang, Doros Polydorou, Owen Noel Newton Fernando, Ryohei Nakatsu & Jeffrey Tzu Kwan Valino Koh (2013). Cultural Robotics: The Culture of Robotics and Robotics in Culture. International Journal of Advanced Robotic Systems,10:400. doi: 10.5772/57260.

Santos Granero Fernando (2007) "Of fear and friendship: Amazonian sociality beyond kinship and affinity" Journal of the Royal Anthropological Institute (N.S.) N.13 1-18

(2000). 'The Sisyphus Syndrome or the Struggle for Conviviality in Native Amazonia' In 'The Anthropology of Love and Anger: the aesthetics of conviviality in native South America,'J. Overing and A. Passes eds. London: Routledge.

(1991). The power of Love: The Moral Use of Knowledge amongst the Amuesha of Central Peru, London: Athlone Press.

Sahlins Marshall 2005. "Hierarchy, Equality, and the sublimation of Anarchy. The Western Illusion of human nature." The Tanner Lectures on Human Values. Paper delivered at The University of Michigan.

Segura, Jordi Vallverdu I. (2011). "The Eastern Construction of the Artificial Mind." Enrahonar: Quaderns de Filosofia 47:171-185.

Turner, J. H. (2011. The problem of emotions in societies (Primera ed.). New York: Routledge.

T. Hashimoto, N., Kato, and H. Kobayashi, "Educational System with the Android Robot SAYA and Field Trial," Proceedings of 2011 IEEE International Conference on Fuzzy Systems, pp. 766–771, 2011

V. Zykov , E. Mytilinaios , M. Desnoyer and H. Lipson "Evolved and designed self-reproducing modular robotics", IEEE Trans. Robot., vol. 23, no. 2, pp.308 -319, 2007.

Viveiros De Castro (1996). Cosmological deixis and Amerindian perspectivism: a view from Amazonia. Journal of the Royal Anthropological Institute (N.S.) 4, 469-88.

Young Philip (1971. Ngawbe: Tradition and Change Among the Western Guaymi of Panama. Illinois Studies in Anthropology No. 7. Urbana: U. of Illinois Press.

Y. Shimon (1999) "Handbook of Industrial Robotics" 2nd Edition. John Wiley & Sons. New York

III

Chemical Aspect of Cognitive Robotics

CHAPTER 3

A learning planet

Hans Sprong

Former team leader of the Philips RoboCup team, The Netherlands.

CONTENTS

3.1   INTRODUCTION

RETURNING from a short trip to France, I found an invitation in my mailbox from my friend Dr. Hooman Samani to write a chapter in his book on cognitive robotics. I was honored and flattered by this invitation, but most of all surprised. Though I have some ideas about this subject and it has remained the focus of my fascination for more than ten years now, I have no claim to any achievements in this field. I am certainly not professionally involved in robotics, artificial intelligence or even machine learning. I am a development engineer at the Philips Healthcare division and as a reliability expert engaged in the development of large medical X- ray devices that are used for minimal invasive procedures in the cardiac and vascular regions of interventional radiology.

The reason why Dr. Samani asked me to write this chapter, despite my less than impressive qualifications, is also the reason why we know each other: Dr. Samani once was team leader of team Persia a participant in the mid-size league of RoboCup. I once was a member and later team leader of the Philips RoboCup team. Dr. Samani and I met in Fukuoka Japan during the 2002 RoboCup world championships. Dr. Samani and his team were in the booth next to ours. I remember our expectations were very high after winning the German Open of 2002, but we did not do so well. In fact we performed disastrous. The robots of team Persia were not very impressive either. In my memory they kept on loosing parts when they moved over the field, like a trailer on a bumpy road, but the team was a very nice neighbor.

The next year we met team Persia again in Paderborn for the German Open of 2003. They had completely new and very nifty robots. They were pretty cleverly made, light weight and simple, but effective. However, because of some of those unexpected troubles that usually plague new technical contraptions, team Persia was kicked- out during the preliminary group competition. We were in the other group and we did not loose a game during this tournament. That is to say, none of the official games, but Dr. Samani invited us for an off the record game early in the morning before the official program would commence. And these little fast moving robots of team Persia, elegantly avoiding our impressive tanks swarming them like mosquito, they beat us convincingly.

Later that year during the RoboCup world championships in Padua Italy we again failed so spectacularly that I felt obliged to resign as team leader. On the other hand, team Persia after fixing the *infant mortality* problems in their robot design became third. They could have won, but they decided to make a small *improvement* in their software before the semi- finals. Not such a good idea usually and they lost these semi- finals. After that they easily won the match for the third place with the previous version of the software restored.

During this championship Dr. Samani, Paul Plöger of the Bonn-Rhine-Sieg University and member of the AIS team and I made some arrangements that would open the way for Dr. Samani to become a student at the university, an intern at Philips and a member of our RoboCup team. Probably that was one of those life- changing events for Dr. Samani; a bifurcation where his career could have taken a different turn all together. He could have finished his studies in Isfahan and possibly become a professor there, but he decided to take another turn.

I also had a life changing moment there during the match we played against team Persia. Needless to say we lost that match, but I witnessed something I had not expected and this enriched my live. I saw emergence right before my eyes. I will return to this in more detail later in paragraph 3 (The Second Law), for now it suffices to say that since then I have been spending much of my free time searching for the reason why I felt so overwhelmed at that moment. This chapter will shed some light on what I have learned during this journey.

But before I start, I need to make the formal statement that none of the content of this chapter in any way reflects the opinions or policies of my employer: Koninklijke Philips N.V. I write this on my own account and I am currently in no way part of any robot related activities within Philips. I also want to express my gratitude to this great and innovative company for giving me the opportunity to experience RoboCup and to explore robotics for a while. Finally, dear reader, if you continue reading, you may wonder whether this chapter is about robots at all, but like a vulture I start in wide circles high above my target, thus setting the stage for the robots to appear and eventually they will. For now, I would like to start as far away from robotics as I possibly can.

3.2 PERSPECTIVE

According to my textbook on astronomy "The Cosmic Perspective[5]," our universe is about 14 milliard[1] years old. In this universe our solar system formed about 4.5 milliard years ago, so our planet is roughly 4.5 milliard years old. There exists evidence life was present already 3.85 milliard years ago. We are a relatively new species and though there are many heated discussions about our origins and which fossils are our ancestors' and which are from other members of the species homo, the oldest remnants may be dated back as far as two hundred thousand years ago. It took a while, but some twelve and a half thousand years ago our ancestors started with agriculture and this laid the foundation for civilization on which our modern society is built.

When we turn our view and look into the future, we may wonder when modern society will end. Some people state it has already ended and we are now in a post- modern society, others give it a few decades or centuries. Some of this will depend on how one defines modern society, but the end will also depend on coincidental catastrophic events that may or may not infer on the vulnerabilities of our society. Then the next question is: whether there will be something we can call civilization after modern society has collapsed? Will the last members of our species live a similar life to our hunter- gatherer ancestors? Will we or others evolve into a new species that is more suited to surviving future circumstances? Then looking further into the future it is clear that somewhere between two to three milliard years from now the sun will have become so hot that no liquid water can exist on this planet[13]. Long before that, human existence would have become impossible and we would have to say goodbye to this *lucky planet* right in the *sweet spot* of our solar system.

I have a vision of a Venus- like planet with all the water of the oceans evaporated to form a dense and suffocating atmosphere depleted of oxygen that has been used to form carbon dioxide with the organic matter that once was living. Though I cannot imagine any carbon chemistry based life to survive under these circumstances, I can imagine robotic life could.

Maybe, if we succeed in creating a habitable atmosphere and an ecosystem on Mars we may hold out over there until our sun explodes in a supernova in about 6 milliard years. Maybe by that time we have figured out how to perform long distance space travel, but I doubt it. I even doubt that we will

---

[1] A milliard is a thousand million often called a billion. However the term billion was originally the naming for a million million (a million to power two). Because the term billion is ambiguous, I prefer the use of the term milliard. The short and the long naming systems are explained in a nice YouTube clip called: "How big is a billion?[10]"

succeed in colonizing Mars or live to see our planet being scorched by the dying sun.

An article in my newspaper[15] inspired by the Maya calendar ending describes four possible catastrophes that could cause the end of our species. One is a prion epidemic. With its long incubation time everybody may have been infected before we even start recognizing the first symptoms. The second is the explosion of a relatively near giant star like Betelgeuse. Such an explosion might X- ray us all to death (with gamma rays, but they are essentially the same). Beside that, we can always rely on the occasional rock from space to crash into our planet. The last big one of about 65 million years ago probably wiped out the dinosaurs. And last but not least there are large quantities of carbon fixed in compounds mixed in the mud on the bottom of our oceans. This carbon was taken out of the atmosphere by plant life over milliards of years, but the mud can be stirred up and the carbon can oxidize to carbon-dioxide. We would not be able to survive a concentration higher than 100 g/m^3.

I have just been scratching the surface of the subjects I mentioned in this paragraph and entire books can be written on each of them, but as this is a chapter in a book on cognitive robotics, just pointing out the vulnerable position of our species entrapped on this planet suffices. Even though I have reached an age where I don't have to worry too much about events that will happen in the future beyond some forty years from now, it still shifts my mood to melancholy when I think of this beautiful planet and its inevitable end, that we as a species are very probably not going to witness.

3.3 THE SECOND LAW

The most important philosophy book published in the twentieth century is without any doubt in my mind the book: "Order out of Chaos[12]" by Prigogine and Stengers. Ilya Prigogine was a Nobel prize winner in the category Chemistry for his work in the field of Thermodynamics. Thermodynamics is by some considered an inferior scientific field; the realm of engineers and not by any means as fundamental as quantum physics. However the originator of quantum physics Albert Einstein is quoted as follows:

"A theory is the more impressive the greater the simplicity of its premises, the more different kinds of things it relates, and the more extended its area of applicability. Therefore the deep impression that classical thermodynamics made upon me. It is the only physical theory of universal content which I am convinced will never be overthrown, within the framework of applicability of its basic concepts."

— Albert Einstein

Thermodynamics was originally based on two laws formulated by Clausius (to which a third was added by Nernst, but that law plays no role in this chapter). The first law of thermodynamics is the law of conservation of energy: "Energy can be converted or transported, but on the whole no energy can ever be gained or lost in a closed system."

The second law for which Clausius used the work of Carnot is the entropy law: "In any closed system the entropy cannot decrease." How entropy can be calculated can be found in any text about thermodynamics, but for my purposes it suffices to say it is often referred to as a measure of chaos. The entropy principle makes heat flow from hot to cold and not in the other direction. So entropy decides in which direction a process will run.

There are many ways in which the second law of thermodynamics can be phrased. For instance: "A perpetual motion machine is impossible," but as a student I learned the phrasing: "In any spontaneous process the entropy increases." This was more or less explained as follows. I have already stated, that heat flows from warm to cold and not the other way, but we are all familiar with a refrigerator, where heat flows from cold to warm, to keep the inside cool. Systems like refrigerators are excellent examples for thermodynamics teachers and as a student I learned to calculate that the engine of the fridge produces enough entropy to compensate for the entropy decrease caused by making the heat flow from cold to warm and some extra so the total entropy increases.

So in any spontaneous process the entropy increases. If the entropy decreases this is not a spontaneous process like a fridge and there must be another process driving it in which the decrease of entropy is generously compensated. As a student I had a misconception of what was meant by the word spontaneous. I thought it meant something like naturally occurring, not initiated by human interference, but from Prigogine and Stengers[12] I learned in this context it means not driven by another process.

As a student I became more and more convinced that not only physical, chemical or biological processes need to comply with the laws of thermodynamics, but that they also govern social, political and economical processes, but what kept me puzzled is how it is possible that while entropy or chaos *wants* to increase one can observe matter ordering and organizing itself into for instance complex living creatures. Only after reading *Order out of Chaos*[12] I learned that while the formulation of the entropy law is correct, it is in fact its complementary formulation we observe all around us: **In an open system the entropy will decrease.** This means that as long as a system is fed with energy or energy is removed from it, its entropy level will be lower than in the static equilibrium state where the energy flow has stopped. Prigogine and Stengers[12] call such systems: "far from equilibrium." Any far from equilibrium system will to a certain extent order itself "spontaneously."

As a reader you may be wondering what thermodynamics and my falla-
cies about it have to do with cognitive robotics. To me that is obvious, but
I understand I will need to explain. Cognitive means something like able to
acquire knowledge or able to learn. Acquiring knowledge and learning are
essentially processes on information connected to intelligence. I don't want
to go into the question what intelligence actually is, maybe here it suffices to
refer to the book: "Understanding Intelligence[11]." I want to go to the level
of information as the subject of the processing. Information and thermodynamics are related. There is Shannon's[7] entropy principle that states that the
more predictable the next symbol in a sequence is, the lower the entropy of
the sequence or the more it is ordered. The less predictable the next character
in a sequence is, the more effort or energy it will cost to find it. If a character is
more predictable, it takes less effort to find it and this in turn means this effort
or energy is already contained in the preceding sequence of characters. This
means this sequence must have more order or less entropy than a sequence
that does not contain information.

I have used the words effort and energy here in a fairly extended context
and not in a thermodynamic or physical context. However, recently a relationship between energy and information has been established, so information
and its processing have become subjected to the realm of the thermodynamic
laws. This scientific breakthrough has killed Maxwell's demon that for a long
time threatened the foundation of thermodynamic theory. On the web a short
BBC film[1] can be found explaining this issue a lot better than I could.

The point I want to make is that the development of life, the biological
evolution and the accumulation of knowledge, first in minds and artifacts, but
currently more and more in electronic data stores, is just one process governed
by the second law of thermodynamics that drives it towards the creation of
more "order" under the beneficial conditions of this planet, unrelenting until
reaching the maximum order or minimal entropy state that the conditions
and the energy flow are able to sustain.

A second takeaway from *Order out of Chaos*[12] for me was the concept of
emergence[2]. While our robots were playing against team Persia in the 2003
RoboCup in Padua, I witnessed it happening. One of our robots got hold of
the ball on our own half. At that time the robots were still allowed to kick the
ball into the opponents goal from their own half. I must say we had a really
hard kick that could easily have gone in. While our robot was slowly taking
aim (the game has become much faster nowadays) the three field players of
team Persia did not do what one would expect, which was to try to tackle
our robot, but instead they lined up to form a wall to shield their goal against
our kick. Now this may seem the obvious thing to do, but it was the first time
I saw this in RoboCup and I must say it looked pretty cool. I was surprised
by this action and I looked at Dr. Samani who was on the opposite side of

---

[2]A very nice book elucidating the idea of emergence is: "The Music of Life[9]," by Denis
Noble.

the field. After loosing the match I asked him about this defence strategy, complimenting him and his team on the good programming, but he told me it was something the robots had come- up with by themselves. Probably they were all programmed to go to the same spot between the goal and our robot with the ball. After the first robot had arrived on that spot the two others just stopped next to it, because the spot they wanted to go to was already occupied.

What I observed here was not a pre- programmed behavior, but an unexpected, but relevant and effective behavior that for that reason looked like an intelligent reaction to the situation. If we can extend Turing's definition of intelligence to RoboCup and state that if you cannot tell the difference between the autonomous behavior of a robot and behavior that is human controlled by a remote, this autonomous behavior can be called intelligent. I found this absolutely fascinating.

3.4   CONTROL OF FIRE

Christmas 2012 I was driving home over the very boring and relatively empty highways of the Netherlands on that evening. I switched on the radio and fell into a program called the marathon interview, in which a prominent person is interviewed for three hours. This time it was professor Johan Goudsblom, a sociologist of whom I thought I had never heard. After about an hour the interview turned to a subject that I could easily relate to. In one of his books: "Vuur en beschaving[4]," *(Fire and civilization)*, he describes the influence the control of fire has had on how the human species has developed. He was struck with an insight while watching the film "Quest for Fire.[6]" When on their way back, after finding fire and escaping from the tribe that held them captive, our prehistoric hero's were attacked by wolves, they defend themselves by throwing burning logs at the wolves. One of the wolf's tails catches fire and the pack runs off. At that moment it occurred to Goudsblom, that having no fur that can catch fire may be a an evolutionary advantage when you live as intimately with fire as humans do. So *The Naked Ape*, as Desmond Morris likes to call us, may have evolved in close relation with fire. On the one hand ridding us of our fur, but on the other hand keeping us warm at will.

After this insight he stated it had taken him another twenty five years to come up with the next one, which is that the human species is the only species that uses dead matter to provide it with energy. This statement may biologically not hold true, but when I heard this on the radio, it occurred to me that we as a species have evolved from a marginal pray animal, running for shelter on the African savannah, to the absolute top predator on this planet's food pyramid because we convert energy into lower entropy in much larger quantities than any other species. It is not so that we are the only species that shape and organize the environment in such a way that we increase our chance of survival. Other species build shelters and decorate the environment with nests, or even entire reefs, but there is no other species that has modified and organized the environment so purposely and extensively as the human species all with the aid of amounts of energy that no other species have available. This is one of those insights that leaves me in a state of utter joy. I received a beautiful Christmas present from Johan Goudsblom who may probably never become aware he gave it to me.

There is still one other aspect of fire that I have found particularly interesting: When I was introduced to systems theory I learned that systems can be described as a number of elements and their interactions. I will need to make a small adjustment to this simple definition later, but lets take this definition for now and look at the question of system complexity. Complexity is a subjective attribute for a system, but there are two dimensions in which the complexity can be expressed. The sheer number of elements is one, the complexity of the relations is the other. A (very anthropocentric) hierarchy of complexity of systems in the relations dimension could look like this:

1. the simple structure where the elements have fixed relations to one another,

2. the dynamic structure or process where elements or their relations can change in time,

3. the controlled process where the process responds to certain conditions or external influences usually to maintain some kind of stability or homeostasis,

4. a simple living system, like a single cell, where homeostasis is combined with procreation,

5. the plant where structures of living cells cooperate to form particular individuals,

6. the animal that is more mobile than a plant and that has more specialized organs for particular tasks,

7. the human being that adds a higher level of intelligence and finally

8. organizations or societies of human beings.

The intriguing question I mentioned before has always been for me: Where is fire in this hierarchy? Is fire a living system? A flame has a structure. Fire is dynamic, adjusts itself to environmental circumstances and it can procreate. I have heard even the ancient Greek philosophers struggled with this question. Of course the next question should then be: If we have robots that can reproduce, would they be living?

Going back to our simple system definition we need three more categories to make the picture complete. The system needs to have a boundary and with that boundary comes an environment that may interact with the system. So we have four categories:

- elements,

- relations between them,

- a system boundary and

- an environment.

Then there is the final less obvious category which is the observer. To me the most amazing is that we have evolved from a bunch of molecules that were able to reproduce themselves into a species that is able to observe itself as a system. Could it be that part of that faculty came from looking into the flames of the fires our early ancestors made to protect themselves against cold and enemies and to prepare food?

3.5   EMERGENCE OF LEARNING

It sometimes looks like a miraculous achievement of a team of engineers when they have succeeded in developing a complex working system like the X- ray machines we develop at Philips, or the robots playing in RoboCup, or for that matter rockets. One may wonder how engineers know how to do this. However, to paraphrase a famous Dutch biologist named Midas Dekkers; the real miracle is that a sperm cell *knows* how to swim to an egg cell to fertilize it and that together they *know* how to build a complete human being from scratch. Is this knowledge? When the first, probably RNA molecular fragments combined and started to reproduce, were they actually learning something? Or were they just obeying the laws of nature for these types of molecules under these circumstances? Or could we say that both are correct observations depending on interpretation?

Imagine you were present at this essential moment in the emergence of life, what would you have observed? Possibly some gooey substance slowly accumulating on a wet surface. Maybe you would ask what kind of detergent from your super market could take that smudge away. Apart from, but closely related to life, something else emerged that is so omnipresent on this planet, but absent in the rest of our solar system and scarce in the universe as far as we can see, something so common to us that it is easily overlooked: Learning. These molecules started to accumulate knowledge of how to reproduce and by reproducing themselves they not only started evolution, they also repro-duced, accumulated and refined the knowledge of how to do just this. They actually created feedback learning.

Can molecules learn? Can chemical substances accumulate knowledge? According to the first definition I could find on the web from the Oxford

Dictionaries, knowledge is: "Facts, information, and skills acquired through experience or education; the theoretical or practical understanding of a subject." One may wonder what the difference is between facts and information. I would say that facts are information, but not all information consists of facts. With skills, I would also argue that we are basically talking about some kind of information, but a person with a skill may not be conscious of the exact information that is possessed, still definitely there is information involved. For instance I can lift my arm at will, but I have no idea how I do this. As a baby I have acquired this skill, but I forgot how I acquired it and now I can preform this movement without even thinking about it. Somewhere in my brain processes on molecular level dictate this action and then it happens. Somewhere in my brain the knowledge of how to lift my arm is stored in the molecular processes that take place inside and between nerve cells.

There are people who believe that our will, our consciousness, our thinking in short: all our mental processes are governed by something outside our body - a spirit - that is not part of the physical world we know. Descartes thought this, but the big question to him was: How is it possible that something that is outside our physical world influences physical processes? That we can move at will if the will is outside the physical world and the movement is inside it? Similarly, there are people who believe that information is a free entity (in the sense of not being bound to another entity) and therefore not subject to the laws of nature. All information we know and even information we suspect to be there, but that we cannot observe yet is contained by substances in our physical world or their properties. (For this statement to be correct, I must include everything we can consider to be consisting of any number of particles - like photons - to be a substance.

The most every day example of a substance containing or carrying information would be a book. But recently, evidence (though not yet proof) for the hyper expansion after the initiation of our universe was found in the polarization of the background radiation. I cannot think of any information that is not contained or bound to something else. Also I would like to point to the relationship between energy, entropy and information as mentioned in paragraph 3 (The Second Law). I have to conclude that my mental processes are physical or chemical processes in my brain.

From the idea that mental processes, like the building- up of knowledge also called learning or cognitive processes, are chemical processes in a brain, it is not such a big step to also call such processes outside a brain cognitive processes. The only difference between the cognitive processes inside a human brain and similar processes outside a human brain is that the human brain is in some cases (but not all) conscious of the learning. Viewed from this perspective it is not so strange to say that relatively soon after its surface started solidifying our young planet started a learning process that has not stopped since.

The learning process of our planet has a relatively simple program, or at least so it seems. The program is to create order or information that is able to survive and expand. Through the milliards of years the planet has become better and better at it. Information is generated, copied and retained or destroyed in inconceivable quantities.

The above statement sounds like a teleological one, as if our solar system has formed to create our planet on which life emerged with the purpose to eventually give birth to a species as advanced, intelligent and creative as we are, a species that has come to a phase in its development where information has become paramount; the crown of creation. However, that is not the way I think of it. What we are observing is an auto- catalytic process, like a chain reaction. Where the chain reaction in an atomic bomb takes only a fraction of a second, our chain reaction is much slower, but it is a chain reaction never the less. Though developments seem slow during ones life, I must say I am amazed when I look back.

When I was in high school I learned how to write computer programs. It was on an IBM 360 and later an IBM 370. I programmed in ALGOL or FORTRAN and I punched my lines of code on punch cards that I had to merge with a deck of machine instructions. The computer was situated in a bombproof concrete building on Leiden's university site and it could be seen through wire- glass windows. It consisted of several large cabinets, some of these were large tape recorders and others, more advanced, were disk stations with the disk drums sitting on top of them. Most cabinets did not easily reveal their function to me, but they had fascinating blinking lights and many push buttons. Half of the large hall was empty. When I asked about it, I was told that room was needed for expansion.

After the expansion had come, in the form of an IBM 370, three quarters of the room were empty. There were times when I would take my bike and ride to this mythical palace almost every day of a vacation and spend most of the day there. Punching cards, bringing them to the DATA 100 to read and then wait and wait until finally on the fanfold form my userID: SGRS130 would appear, so I could start trying to interpret my error codes. Now I have a smart phone that I carry around in the pocket of my shirt that has the processing power of a multiple of these old IBM 370s.

In retrospect it looks like there is little relation between an old main frame computer and a smart phone. They are completely different things in appearance and usage, but inside they are just processing data. It is obvious that the smart phone is much more efficient at it. An interesting paradox is that the more efficient data processing becomes, the more need we appear to have for it and the need still grows faster than the availability. There seems to be an unlimited need for data processing, or at least from where we are now there is no end in sight.

Even though the data processing becomes more and more efficient, energy is still needed to make all this processing possible. Recently I saw some pictures of the chillers needed to keep the server park of a data center cool. They are impressive and so is the amount of energy involved in generating, maintaining, storing and distributing all these data that keep accumulating.

On the total energy consumption of this planet, data processing probably still takes a relatively small part, but I would not be surprised if it is the fastest growing part. It will eventually become the largest energy sink, if my interpretation of the evolutionary process is correct.

There is just one more thing I would like to mention before finally turning to robots: The European community is sponsoring one big science project called: "The Human Brain Project." It will cost several milliard Euro and the idea is to build a computer that is able to simulate the human brain. If I am correctly informed by my newspaper the computer will be the largest computer ever and it will consume an amount of energy that is equivalent to the consumption of a small town. You can imagine especially this last bit of information was interesting to me. The effort goes into simulating something we have all gotten for free, that fits inside a small fish bowl and that consumes about 200 Watt of power. This makes me wonder about the assumptions on which this project is based. The American government is adding to this a project to measure and map brain activity. Possibly the European computer can be fed with the data the Americans collect. Apparently Google is also working on a computer of its own. After building the program that is able to win Jeopardy by just *reading* all kinds of random texts, it is now trying to build a "simple" model of the human brain and Ray Kurzweil is involved in this. I'd put my chips on Google and Kurzweil, because I like simple models. A simple model can often suffice:

"Essentially, all models are wrong, but some models are useful."

Geoge E.P. Box and N.R. Draper
from WikiQuote

I don't have any doubts that computers or robots can be intelligent, that they are intelligent already, but I am not so certain that we can ever make them think like a human being, let alone that we can actually simulate a particular human being in a computer. The reason for that is that apart from the structure of the brain, the genetic information, the epigenetic environment, the hormonal system and the sensory and actuation systems need to be all perfectly simulated and tuned but once the simulation has started, the simulated brain needs to observe and feel exactly the same things the "real" brain would have, not to diverge. However, for the evolutionary process this will be of little importance. What will be important is: will the computer or the robot brain be able to compete with other species and prevail?

Interestingly for this question, I have just read a hypothesis in my indispensable newspaper[3] that the reason Homo sapiens prevailed over the Neanderthal species maybe because of their more efficient learning. They were not that much smarter, apparently the brain volume of the Neanderthal was slightly larger than that of the Homo sapiens, though brain volume only correlates lightly with intelligence. The hypothesis put forward by professor Joseph Henrich of the University of British Columbia and professor Mark Thomas of the University College London, claims in short that sufficient individuals in a group should be able to teach the younger individuals to maintain and expand the knowledge of that group and in this way support its technological progress. Neanderthal men did not seem to survive long enough to pass on what they had learned, possibly because of their dangerous hunting habits. Also there seems to be no evidence of Neanderthal group gatherings around a camp fire as there is for Homo sapiens. Arguably the offspring of the early Homo sapiens had an evolutionary advantage because they could sit at a camp fire listening to the teachings of their grandparents.

3.6   OF MEN AND CENTAURS

While I was a member of the Philips Robocup Team I spent some time working on a project to explore the possibilities for home robots. One activity in this project was to visualize such a home robot. For this we asked a student from a design academy named Leonie Hurkx to help. As an intern she made beautiful designs and considered many alternatives for skin colors, light effects and she even came up with the original idea not to give the robot hands, but to give it tentacles like an octopus with suction cups decreasing in size towards the tips. The project gradually tapered off and I think most of what we did was forgotten, but I still think there were some good ideas in it.

One idea was to give the robot a Centaur configuration and make it as big as a large dog. The main reasons for this configuration were: that only a legged robot would convincingly be able to operate in a home environment without major adaptations to this environment. A four legged robot would be a convincing beast of burden, but it would also need hands or, as Leonie defended, tentacles to pick- up and operate tools made for human beings. Centaurs have no negative connotation like spiders with six limbs and last but not least in a Centaur configuration one can fold- up a relatively big length while the head of the robot could remain relatively low, so even a relatively small human being would stand a bit taller than the robot who is his or her servant while this robot could fold out to reach for high places.

I think the robots will develop initially as helpers for elderly and disabled people and a lot of effort indeed already goes into this direction. Once a robot has been developed that can autonomously take care of people that can not take care of themselves and their households, they only need to be produced in larger quantities to become generally affordable household robots. The implications of an autonomous household robot that can actually perform virtually all household tasks, are manifold but the most obvious is that there is no reason why such a robot would not be able to perform most jobs, including *manufacturing robots.*

From general household robots, to robots that can eventually perform any job human beings could is only a small step. Such robots would be able to reproduce. When we have entered this stage, the robots have become a new species or even a new domain of life. One of the most interesting current developments is not so much in robotics, but in what is now commonly known as 3D- printing. Even though this is already a pretty sexy subject, one may think of 3D- printers as the reproductive organs of the future robots to make it even more sexy. The possibilities of 3D- printers are already amazing in making all kinds of possible or impossible constructive elements, but think of what can happen if substances containing nano machines can be printed to construct muscle- like or nerve- like structures intricately interwoven with bone like structures or joints.

One of the learning points when working with robots is that what we find easy as humans can be very difficult for a robot. Simple things like understanding and speaking a native language, something you learn naturally, recognizing colors under different lighting conditions, recognizing faces, recognizing the smell of substances or even tactile feedback. In all of these fields robots are still inferior to us, but they are making progress and the interesting thing about robots is probably that though they learn very slowly compared to humans, they learn very efficient, because what was learned by one robot can easily be copied to the next.

Some are expecting major break- through developments in a few decades, but I have become a bit more tempered in my expectations. Ray Kurzweil[8] predicts what he calls the *singularity* will happen in the year 2045. According to Kurzweil in 2045 there will be a processor that has the capability not just to emulate the brain of a single person, but all the brains of all people in the world. This may be true, but if this processing power is not used for learning, if it is still depending on what we as humans feed into it and ask it to do, this processor will probably be busy mostly with what we as humans want. To know what we want you just have to look on the internet.

Apart from very useful applications like Wikipedia most of the internet is filled with trivia about the personal lives of people who use the social media and I read somewhere that pornography takes the largest part. So even if we have succeeded in creating a computer that is as clever as all human beings together, it will probably be occupied with creating the ultimate satisfying personal sexual experience; one that you can share with whoever you would like to share it with. Imagine how many *likes* that would get. Furthermore, in the light of recent revelations, the rest of its time it will probably be busy nosing in everybody's private communication trying to find threats to society. Probably by that time the predictive capabilities[3] have become so convincing,

---

[3]Chaos theory makes it unlikely that such predictions could actually be accurate, but given the fact that many people would love to be able to predict the behavior of others and the fact that many people rather believe in what they want to exist than in what science has to offer, it can be expected that such behavior predictions will be attempted.

that you can be arrested before you have even thought of a crime you could have wanted to commit.

Only if the processor becomes independent, it can start to learn autonomously to make use of its intelligence to the fullest and accumulate new knowledge. For this the processor needs to be a robot brain. It is not necessary that the brain is carried around by the robot. Robot and brain may well exist separated, but there cannot be intelligence without a body. This credo of the synthetic method is nicely illustrated in Valentino Braitenberg's book: "Vehicles[2]."

Scientists from various disciplines predict or warn against the oppression of humans by robots or computers once they have become more intelligent than us and they all think this moment is nearing. I have the impression this is partly based on confusing mere processing power with intelligence. I have the idea that it is not intelligence, but learning capability and efficiency that will eventually prove to be the essential property that will decide this competition. This will place the point when the robots pass us a bit further in the future, but it will on the other hand make that point more inevitable.

For the versatile robots to appear, that can indeed reproduce themselves autonomously and that may claim to be a new form of life, not only the developments in the field of processing power are needed. Many technologies need to come together in the field of sensing, materials, remote communication and mechanics of walking just to name a few, but most of all, there needs to be a common idea that everybody needs a robot companion like in the early days of personal computers, when there were programs popping-up everywhere to provide employees, students, school kids, elderly people and anybody else with a PC that was mostly useless and that after causing a lot of excitement and fuzz when it was first taken out of its cardboard box after a few weeks would be stored in the attic collecting dust. Remember how expensive such computers were?

What will the future bring? I think there are three possible scenarios:

1. We don't succeed in developing a robotic species.

2. We merge with the robots as Kevin Warwick[16] advocates as a way to survive.

3. We create a robot species and we loose control over them.

If we don't succeed in creating a robot species, we will probably perish because of some global disaster or maybe even because of our ignorant behavior with this fragile planet. Whatever disaster will happen, I cannot imagine that it will destroy all life on this planet, so after we have gone, evolution can take over again and give it another try.

On the other hand, if we succeed in creating a robotic species, it will have many consequences. In the first place the economy will be destroyed, because for most people that have an income, they earn it by selling their labor. If labor has become something for your robot to do, how are you going to earn money? Especially if you don't have a robot. Then the robots will by the entropy law that was described earlier expand and start competing with humans for resources; in the first place energy, but carbon may also become a scarce resource. However, considering the composition of the earth's crust, it is likely that the robots will be predominantly made out of silicon and iron.

An if-you-can't-beat-them-join-them- scenario in the most literal sense of the word is described by Kevin Warwick who sees a development of men into cyborgs as inevitable. Currently in the development of wearable electronics and even more so in the development of prostheses that connect to the nerve system one can see evidence of the viability of his view. However, when the cyborg develops to become more and more machine and less human and if Ray Kurzweil is right about the brain easily being replaced by a much better chip from 2045 onward, why would a robot still want a vulnerable human part?

Consider a robot that is able to provide itself with energy and that is able to reproduce itself and then ask yourself the question: Which part of me would give this robot an evolutionary advantage over another robot that does not have such a part? To be frank, I cannot think of any part of my body that a robot would be particularly interested in considering also that it will have to sustain it.

To dwell a bit longer on this idea, let us consider the following: According to some evolutionary biologists starting with Charles Darwin the human brain must have developed under the pressure of sexual selection. The typical traits of the human brain like creativity, appreciation of beauty, curiosity, abstract and symbolic thinking, they were all adding to the attractiveness of sexual partners and at the same time gave us as a species some small advantage over competing species in acquiring energy. Initially this was in the shape of food, but later on, after we learned to control fire, we opened a niche with unlimited potential where we are still the only inhabitants. Because of the abundance of energy and the lack of competition from other species we never have become very efficient in its usage.

To speak about efficiency one has to first set a goal or have a purpose and then consider the cost at which this goal can be reached or the purpose can be fulfilled. If we take the purpose stated in the previous paragraph: "to create order or information that is able to survive and expand," we only use a very small part of our total energy consumption for this purpose. We use a lot of energy to sustain a body that consumes almost as much energy when doing absolutely nothing as when it is working, thinking or learning. Then we waste enormous amounts of time and energy on our selection and mating processes. We have to learn everything by reading books or by what is called learning it the hard way - by experience (feedback learning) - and after we have accumulated all the knowledge we can rake together, ... we die. Learning can be done much more efficient by a species that can switch off its energy consumption when it does not need to do anything, that does not have to find a mate to reproduce, that does not need to display its attractiveness to be found and that after its body has worn out can be transferred into a new body without any loss of knowledge. A species that will be able to communicate among one another with full two directional telepathy, that needs no sleep, that can work while being charged, that can consult a super computer on any question it may have at the blink of an eye. Creatures that may occupy bodies prepared for them on the other side of the planet or on the moon or maybe on Venus or Mars, just by transmitting the data from one body to another.

Once we have created our centaur servants, they will become competitors in our niche. Initially they will be clumsy and depending on our service in maintaining and feeding them, but slowly they will become smarter and more efficient. Once they have learned to reproduce themselves and to control the undoubtedly complex logistic chain of the parts and materials they are made of, they will also find out what limits there are to their expansion and they will learn how to overcome those obstacles. Possibly men may for a while live in a paradise like situation as these robots were originally developed to serve men. A situation may arise where there is an abundance of everything and nobody will have to work. However, I have no doubts that people if they have nothing else to do, will find many ways to make one another very unhappy, because even if there is enough of everything, people will still be greedy and ignorant and our brains are not made for sharing abundance we are still programmed to function in an environment of scarce resources and competition.

Once the centaurs have gained control over their entire production process and are completely independent of men, our species will have become irrelevant to them and probably they will treat us the same way we treat other species. I don't think the robots will be as cruel as people are, they will basically be just as thoughtless about us as we are about the beings and things we find irrelevant.

Still the question is not so much if we will eventually perish as a species, because we will, but if we can create a species that has a better chance of survival and I think this is quite possible. If we step back to the short list of disasters that may destroy human life in paragraph 2 (Perspective): a prion epidemic, a giant star explosion, a large impact from space or the carbon released from the ocean floor, I think the first and the last would hardly affect the robots at all and the other two may have an impact on the robots. Possibly the electro- magnetic pulse from the star explosion could affect the electronics, but I think robots would have a much better chance of surviving. Of course, robots will have their own vulnerabilities. They may depend on specific rare materials that are only found in specific locations on this planet, and such mining locations may be destroyed. Or a computer virus may spread and make their processors useless, but there are ways to protect yourself against such vulnerabilities. Remember robots will be very disciplined and they will not open a dubious site because they are tempted by the prospective of seeing a picture of a good looking girl, a funny cat or winning a lottery.

3.7   GREAT ESCAPE

After his internship with our RoboCup team Dr. Samani left to try his luck on the other side of the North sea and from there he moved on to the University of Singapore. In 2008 we met again during the First International Conference on Human Robot Personal Relationship in Maastricht. Dr. Samani gave a lecture on Lovotics: the combination of love and robotics; the subject of his PhD thesis [14] .

On Friday June 13th, the first keynote address of that day was given by Dr. Anne Foerst of the Saint Bonaventure University, New York titled: "The Community of Human and Non-Human Persons." I was intrigued because her topic was the combination of theology and robotics, a combination that for me is unimaginable. I was also intrigued by at least one thing she said. (I have to paraphrase, from memory, because I did not take notes and I could not find proceedings.) If I remember correctly she said that the people most prominent in the development of robots were driven by a religious motive. In an *Imitatio Dei* they wanted to follow the example of God and become creators themselves. I found this very interesting, because most of the time I haven't a clue about my motives for doing things myself, let alone that I understand the motives of others. However, her point of view inspired me to end this chapter on a more or less religious note.

When I was in high school I became interested in Buddhism. I learned the Buddha was a prince who's father had taken provisions to shelter his son from any sorrow or harm. When the young prince finally escaped from his golden cage, he came into contact with an ill person, an old person and a corpse. From these meetings he concluded that life equals suffering. At that time that was a bit of a big leap for me to accept. Even though I was not such a happy kid in high school, I was convinced my time would come and it has. Still, I have, much like the young prince, been confronted with enough illness, despair of old age and death to understand suffering a bit.

In the previous paragraph I have entered a paradox, that I will expand here. First I stated that there is an abundance of energy and then I state that the robots will compete with us for scarce energy. So is there an abundance or a shortage of energy? Looking at the sheer amount of energy in the universe, there is an abundance, however most of it is not available to us. When we started to use fire as a new source of energy, something other than food, but still in direct relation with food either to prepare it or as a hunting aid or even to survive in colder climates where more game animals were found, food was probably still the limiting factor for the expansion of the species, and the usage of fire was only a small evolutionary advantage. Now we can live on almost any surface of this planet and we can even survive for a long time in space thanks to the newly discovered sources of energy and the artifacts and food that we can produce using this energy. As a species we were able to expand through a further expansion of our energy usage and the discovery and exploitation of more and more sources. It seems a law of nature that any species will expand until it reaches the limits of the resources sustaining it. This means that any abundance is usually gradually depleted by a growth of the population. And in any population there will also be a distribution of wealth and for this reason there will be individuals in the population that are doing reasonably well whereas others will live on the edge of starvation or deprivation. And though this observation is, as far as I can see, true for all living creatures, it is even more true for humans, who have found ingenious ways for storing and passing on wealth from one individual to another. Where a "wealthy" mammal can store its wealth in body fat to a certain extent, we can store our wealth on a bank and give it to our children. But that is not the point I want to make. What I want to point out is that the growth of a population and the stress this causes on the resources sustaining it will cause a number of individuals in the population to live a marginal life of suffering.

Once robots master the entire logistic chain of their own production, they will like any species not stop when there are enough robots, they will also (like we have done before them) deplete this planet's resources until they are stopped by its limitations. Possibly they will approach this point a bit more cautiously than us, because they will probably be much better at thinking about consequences of their own actions especially on the longer term. They will also see that inhabiting only one small planet will make them vulnerable to cosmic catastrophes. They will expand to neighbouring planets and beyond.

One major question for me is if robots will have emotions. I think we may be able to give robots emotional behavior that cannot be distinguished from the emotional behavior that we have or recognize as such. Dr. Samani [14] has demonstrated that he can make a robot that behaves like a loving pet. The next question is however, if, once the robots have taken control over their own production and design, will they preserve this behavior, or will it gradually be weeded out as superfluous by the robotic evolution? Will there be a robot to shed a tear when the last human being departs from this planet? For some reason I have the feeling that for robotic survival and expansion our passions may not be needed.

Aristotle in his Ethics states that happiness is the highest human goal to strive for. The reason for that is that you don't have to ask: "Why do you want to be happy?" It is evident to him, that happiness is the ultimate goal. But if I ask myself the question why would happiness have that status, the only answer I can think of is that everything that makes me happy may have something to do with my preservation and procreation. After making certain my body is in good shape the next thing I want to do is to demonstrate my attractiveness as a sexual partner through whatever I can think of to make myself notable. So while we are all trying to do our best to reach the milestones on the path to happiness for our own sake, we are in fact expanding our species and our knowledge as little gears in an ever growing machine programmed to learn. Happiness and the avoidance of suffering are on the level of our perception the driving forces to make each individual contribute. For the robots, as they do not procreate through sexual intercourse between two individuals, the natural selection process will be completely different. They will probably develop a completely different selection mechanism. I am pretty certain that suffering will not be part of that mechanism and I have my doubts about happiness.

For the planet's learning project to continue, there must be one drive incorporated in the robotic mind on a level that we might call an emotion and that is curiosity. Only if the robots are curious about things, they will ask questions, explore and find answers. They will certainly need to find ways to make them less vulnerable to the dangers that threaten them. Though none of the four threats that could wipe out mankind as mentioned in paragraph 2 (Perspective) would have the same effect on robots, robots will have vulnerabilities of their own. Whatever these vulnerabilities will appear to be and in whichever way they will succeed to cope with them, eventually to sustain the species, they will need to leave the solar system and find a home elsewhere. Maybe they will find ways in which they can travel to planets outside our solar system. Maybe they will find a way to direct *space ship earth* to another star. And maybe - a bit like a Craig Venter prophecy - during their exploration of the Galaxy, they stumble upon a planet on which something like human life would be sustainable and maybe, just out of curiosity, they will take the data of how to build a human being out of the ancient archives and re- erect this troublesome species that is not only aware of its own mortality, but that has also become aware of its inevitable extinction.

Be gentle to the robots, for they will inherit this world.

3.8   Bibliography

[1] Jim Al-Khalili. Maxwell's demon and the nature of information - http://www.disclose.tv/action/viewvideo/154789/ maxwells_demon_and_the_nature_of_information/.

[2] Valentino Braitenberg. *Vehicles: Experiments in Synthetic Psychology*. MIT Press, 1986.

[3] Lucas Brouwers. De macht van het getal - (the power of the number). *NRC Handelsblad*, 2014 Apr 19.

[4] *Fire and Civilization*. By Johan Goudsblom. London: Penguin, 1992.

[5] Nicholas Schneider Jeffrey Bennet, Megan Donahue and Mark Voit. *The Cosmic Perspective*. Pearson/Addison Wesley, fifth edition, San Fransisco, 2008.

[6] Gérard Brach (screenplay) J.H. Rosny Sr (novel) and Jean-Jacques Annaud (director). La Guerre du Feu, Canada, 1981 - (quest for fire).

[7] George J. Klir. *Uncertainty and Information: Foundations of Generalized Information Theory - Chapter 3: Classical Probability Based Uncertainty Theory*. John Wiley & Sons, 2006.

[8] Ray Kurzweil. *The Singularity is Near*. Viking, New York, 2005.

[9] Denis Noble. *The Music of Life: Biology Beyond the Genome*. Oxford University Press, New York, 2006.

[10] Numberphile. How big is a billion? - http://www.youtube.com/watch?v=c-52ai_ojyq.

[11] Rolf Pfeifer and Christian Scheier. *Understanding Intelligence*. MIT Press, 1999.

[12] Ilya Progogine and Isabelle Stengers. *Order out of Chaos: Man's New Dialogue with Nature*. Bantam Books, New York, Apr 1984.

[13] Osborne Rushby, Claire and Watson. Habitable zone lifetimes of exoplanets around main sequence stars. *Journal of Astrobiology*, 2013 Sep 19.

[14] Hooman A. Samani. *Lovotics*. Lambert Academic Publishing, Saarbrücken (Germany), 2012.

[15] Alex Tieleman. Vier superrampen die de mensheid kunnen vernietigen - (four catastrophes that can destroy humanity). *NRC Handelsblad*, 2012 Dec 20.

[16] Kevin Warwick. Kevin warwick - home page - http://www.kevinwarwick.com/.

IV

Physical Aspect of Cognitive Robotics

CHAPTER 4

Embodiment in Cognitive Robotics: An Update

Paolo Barattini

Ridgeback SAS, Italy

CONTENTS

4.1 INTRODUCTION

IN the present chapter we will deal with the recent implementations of cognitive embodied controls in robots that are realizing in an embrional way the theoretical stances fostering the introduction of advanced features as lower level sensory motory embodiment, intentions, motivations, self generated goals and emotions. The wording "embodied cognition" it is some decades old. Actually the first use of this label is not exactly known. It is used by authors with different meanings, essentially to express the hypothesis that the human cognition (whatever it is) is not a process that takes place inside the brain thanks to logical and computing functions and neuronal circuitry that elaborates external information and produces an output. Embodied cognition foresees that higher level reasoning is influenced or even concepts and symbols are created from bodily derived experiences and phenomena. For example from sensory-motor input.

The philosophy background and discussion about these different views is as old as Plato. More recently such positions have been defined as

Cartesian versus Heideggerian [1]. The concept of embodied cognition arose in two cognate scientific areas, psychology and computer science including valuable contributions from the robotics branch. The wider use of the term embodiment and an enlarged embodied perspective have been in the course of their historical development compounded thanks to contributions from sundry areas like Neurosciences and Biology. In each area the wording embodiment appear to have been used with slightly different meaning. Relevant historical contributions includes the works Gibson [23], Damasio [13], Varela [19], Dreyfus [17] and many others.

The embodiment concept in Artificial Intelligence and robotics is mainly an English speaking countries science history. The embodied cognition concept arose in the psychological field as the evolution and reaction to the mainstream psychological current of the cognitive psychology. In the computer science it appeared as the reaction to the difficulties encountered in the field of artificial systems and expert systems, representing the next step in the evolution differentiating from the GOFAI Good Old Fashioned Artificial Intelligence. [1]. The technical progress and the adoption of new concepts related to the embodiment in computer science and robotics is interlaced with the scientific and philosophical debate about possibility for a machine to display intelligence or consciousness, the related mind body problem and sundry similar issues.

Recently Robbins and Aydede [5] provided a wide range panoramic view of related issues presenting the situated cognition wording as the more general category. They classify other concepts related to embodied cognition, such as embedded cognition, extended cognition, enactive cognition, distributed cognition, as sub categories of the general term Situated Cognition. A recent account of these terms is proposed by Ward et al [41] that additionally include the Affective dimension, the point of view that cognition is influenced by the evaluative stance of the cognizer. Actually the most recent steps in the embodied cognition approach try to include the emotional, affective, and intentional dimensions in Robot controls modelled according to this technical and theoretical stance. In other words the basic stance that is should be possible to improve the robot performance and intelligence widening the embodiment so to have an organismic approach including much more aspects of living intelligent organisms.

In recent years several authors have argued that a truly biologically inspired and truly cognitive robotics would need to go beyond sensorimotor interaction in robotic models of embodied cognition by also taking into account homeostatic/emotional dynamics, i.e. the interplay between constitutive and interactive aspects. In other words including sensory-motion embodiment contributions to the robot behaviour provides some improvements but it is still faraway from an intelligent adaptative behaviour that includes intentional states and the affective dimension as part of the loop generating intelligent self-adaptative behaviour. Indeed the ongoing research about embodiment aims to go beyond the sensory motor level, field in which there are successful completions in robot movement control driven by the lower level components. Di Paolo [16] relates the current developments to the philosophical theoretical stance of Dreyfus [17] that criticised the artificial intelligence with regards to the use of modelling the human mind as information processing.

4.2 TALKING CHINESE, MEANING AND RECENT THEORETICAL DEBATE

One relevant problem in artificial agents and robots is that of the "meaning". The "meaning" is the missing tile in an approach considering information processing (the traditional cognitive approach) as well as the later approach including "embodiment". In the words of Di Paolo [16]: "the body, not some algorithm, is the originating locus of intelligent, concernful activity (if by intelligent we understand intentional, directed and flexible), and the world is not the sum total of all available facts, but the world-as-it-is-for-this-body." The problem of the intentional states and related issues of "understanding" and in an extended way "meaning" has been exemplified by the Chinese Room case devised by the philosopher J.R. Searle [39].

The reknown imaginated experiment foresees a men closed in a room. He does not speak or understand Chinese. He receives two groups of Chinese characters. He receives also rules that connect characters from one group to characters of the other group. The men becomes proficient in using these rules so that the results of his activity mimicry the capacity of a native Chinese speaker. If he receives a third group of characters from outside the room (these are questions about the previous set of characters) and he applies the rules to connect them to the previous characters and then sends out the correlated characters, he will be providing answers. These will appear to Chinese speakers indistinguishable from those of another person proficient in Chinese. Nevertheless the person that is in the room does not understand Chinese.

If the human that speaks English is in the same situation (answering questions about a story in English) he does not need the set of formal rules to correlate characters. The Chinese character correlation can be implemented as a computer program. So were is the difference between the Chinese case and the English case? In the English case the man understands that the story is about a man, a restaurant and eating hamburgers. In Searle's words "understanding implies both the possession of mental (intentional) states and the truth (validity, success) of these states".

Searle arguments: "It is not because I am the instantiation of a computer program that I am able to understand English and have other forms of intentionality (I am, I suppose, the instantiation of any number of computer programs), but as far as we know it is because I am a certain sort of organism with a certain biological (i.e. chemical and physical) structure, and this structure, under certain conditions, is causally capable of producing perception, action, understanding, learning, and other intentional phenomena. And part of the point of the present argument is that only something that had those causal powers could have that intentionality." So the problem of intentional states is, in his words, related to embodiment.

Searle's case is closely related to the Symbol Grounding Problem (SGP). Essentially how can we connect objects of the external world to words (symbols) that are in our head? How can symbols be invested of meaning? This is a challenge in human psychology, phylosophy but also for robotics research. And if the meaning is grounded by embodiment then also internal states and emotions can influence the meaning and have influence on motivations and intentional states that, at least in the human, appear to influence goals and self determination compounding the cognitive, rational, and logical aspects. The Symbol Grounding Problem in robotics been recently object of a review by Coradeschi [12]. She considers two categories: the category of physical grounding meaning the creation of connections between symbols, where these can be categories, such as shapes, spatial features, colours etc acquired through the robot sensors from the physical world. Within the physical grounding she introduces the subcase of perceptual anchoring i.e. to connect the sensory data to higher level conceptual symbols; and the category of symbol social grounding that is after robots has been able to ground the sensory data to symbols, how can multiple robots share these grounded symbols? The research in robotics and AI on one side provides new models that can possibly help to explain how the human intelligence and cognition work, on the other side robotics is finding inspiration in the living organisms.

The latter line of research is therefore confronted with the definition of some basic organismic features. DiPaolo [16] introduced some refinement in the basic idea of how to characterize the organism beyond homeostasis preservation so to be more amenable to agency and sense making through the internal events and loops that underlay the sensory motor interaction with the environment and the sense-making of living organism. He adopted the concept of habits referring to some antecedents like Ashby [4], Goldstein [24], Varela [40], and Piaget [35]. He tells "Finding food is necessary for survival, but often there are different sources of food and different ways of obtaining it. So possibly there are preferred behaviours, and consequently we can speak of habits and also of the way in which habits come about. A living organism as well as a robot needs to preserve its life but possibly there are multiple choices to achieve survival. So far instead of life we can speak of way of life.

Habits supporting the way of life are built out of networks of processes. The habitual patterns of behaviour are capable of adaption and this takes places when it is challenged the process of formation of habits (so not necessarily the organism survival). Eventually the robots shall be provided not with life but with the capacity to generate and have ways of life." In his view the mechanisms that allow the acquisition of a way of life will also generate intentionality in that they orient the robot to choose one way of life in alternative to other possible ways of life, under the condition that they are compatible with survival.

On top of this he adds that a corollary is the need of plasticity, the possibility to change when challenged by the environment or to preserve a habits. Apart intentional agency, the approaches referring to biological knowledge and to living organism may include autonomy, internal sensory input (for example the equivalent of signals, sensations form the viscera), emotions and their autonomous nervous system, hormonal and behavioural correlates as part of a wider view of embodiment.

Ziemke [42] reviewed and widely discussed the landscape of robots embodiment. He also highlighted the basic research issue, that many studies and discussion about embodiment skip which kind of body shall a robot have. He identifies five possible categories of embodiment:

- structural coupling between agent environment,

- historical embodiment as the result of a history of structural coupling,

- physical embodiment,

- organismoid embodiment, i.e. organism-like bodily form (e.g., humanoid robots),

- organismic embodiment of autopoietic, living systems.

This is some kind of hierarchical structure from the more general category of structural coupling that it is applicable also to a geological system with meteorological interaction to the most restrictive category of organismic embodiment. The latter is a possible faraway target for robotics research that currently is more of the organismoid type.

Actually in the scientific literature there are some research streams that descend from different theoretical as well as technical stances. For example we have the connectionist approach that is more or less inspired by studies on the computational aspects of cognition and consciousness in intelligent devices [29]. Clark [11] considers three types of connectionism. He adds a fourth type to describe the newest developments in the field.

The so called First-generation connectionism that began with the perceptron and the work of the cyberneticists in the 50s. It involved simple neural structures with limited capabilities. Their limitations draw criticism by representatives of the symbolist AI school in the 60s, which resulted in abandonment of connectionist principles by mainstream researchers for decades. The Second-generation connectionism: the connectionist approach was revamped in the 80s. It extended first-generation neural networks to model complex dynamics of spatiotemporal events. It developed architectures based on neural networks and learning algorithms.

The Third-generation connectionismor dynamic connectionism: the definition is ralted to the research of the 90s that studied and modelled more complex dynamics sometimes inspired by biological knowledge producing for example Distributed Adaptive Control models. The Fourth generation connectionism: involves chaotic dynamical systems to model data derived from the EEG analysis of human brain structures.

4.3 RECENT PRACTICAL IMPLEMENTATIONS OF THE EMBODIMENT OF EMOTIONS, MOTIVATIONS AND INTENTIONAL STATES

Regarding the inclusion of the affective states and of emotion as part of the robot control and interaction system, there have been some research experiences that implement some of the concepts reviewed in the previous paragraphs. Each of the practical embodiment, i.e. each of the implementations in a real world robotic system refers to sundry different definitions of what emotions, motivations and intentional states are.

It must be pointed out that the area of emotions, motivations and intentional states is a theoretical quagmire of different definitions and approaches. The background psychophysiological, neuroscientific and psychological knowledge itself is a quagmire. Actually is more human constructed and less natural than the robot itself. For example is widely debated if emotions are natural kinds [38, 18, 25].

This simply means that it is widely discussed if the words we use corresponds to natural phenomena and entities or if the words are more a scientific or social construct. As a matter of fact there is no general agreement of how many emotions there are (simply consider how many and which words identify emotions); furthermore sometimes emotions are classified (or better the word we use to describe emotions) in primary (or basic) and secondary. The science, i.e. the scientific investigation of emotions using objective measurements do not help much. many different physiological variables can be measured (heart beat, galvanic skin reflex, respiratory rate, eye blink, EMG, EECG and MR scans) but this does not provide a unique standardised classification nor help to characterise univocally an emotion. Furthermore some of these variables are also considered indexes useful to measure cognitive loads. Additionally we have moods (long term affective states) and emotion appraisal, as well as valence (positive or negative) of emotions.

Consequently each Robotics scientist that implemented and embodied emotions in an actual robot has been confronted with the need to choose one of the possible stance about emotions, one possible group of basic emotion and one possible concept of how emotions are related to behaviour. Additionally experiences have been implemented on different robotics platforms and in the frame of different scenarios and applications. This of course makes it difficult to compare results and to assess if the embodiment of emotion brought about an improvement in robot performances. Furthermore sundry robotics system and platforms have been used by different researchers. No benchmarking appears to be possible at present.

Nevertheless these shortcomings the research has progressed and provides many interesting experiences. Emotions apart, if we go deep in the background definitions regarding motivation, intentional states, and agency, we will again be confronted by a host of definitions, approaches and experiments providing a lively but wavy background.

Additionally there are sundry areas of research involving robot autonomy, emotions, motivations and intentional states: the improvement of robot behaviour in dynamic environments (a more organismic intelligent robot), the improvement in human robot interaction, i.e. a robot that is able to express affective states in a way appropriate to the situation and comprehensible by the human, in some way the robot should give the illusion of life [36], the developmental learning inspired from studies of human child development. The latter and also the human robot interaction research line, sometimes called affective computing, are in some way independent from that of embodied autonomous intelligence which is more concerned with some basic features os sensory motor behaviour often referring to the sensing, acting, perceiving, adapting principle and behavioural loop. We will present here mainly the recent achievements in robotics embodiments bringing about autonomous behaviour including self generated goals or motivation and sometimes exploiting also emotions. We will not consider the so-called social robotics.

Choxi [27] implemented a system based on a model in which each emotion level is treated as a continuous property of the system. This includes "emotional persistence", that means that the emotion levels are controlled in order to resist to change. Additionally they included machine learning techniques.. They tested their technology in a human robot interaction scenario, in which the robot moves in the interiors of a building. The mission of the robot is to reach pre-specified location where a certain person is located. To achieve this goal the robot has to interact verbally with people in order to receive help and to collect information. Furthermore it has to avoid many obstacles during navigation. The robot emotions are useful for the interaction between humans and the robot. Apart the emotions the robot includes a so called motivation engine. The use for of the motivation engine is to improve decision making in relation to the choices needed navigation.

The motivation engine receives input form external perceptions and also receives status messages originating from its speech and behaviour systems. The motivation engine produces values are handed over to the behaviour selection system, which on the basis of the motivation values and of the current state of the robot will choose the next action to be taken. These are enacted by the speech and behaviour module, that controls the lower-level components of the robot. The motivation engine, is based on four basic items: frustration, friendliness, fear and fatigue. Friendliness is set to increase when there interactions with humans. The frustration builds up when the robot does not proceeds in his mission. Fatigue is measured as the energy level of the robot; an increase in fatigue will trigger the low energy mode and related behaviours.

Manzotti [30] provides some basic definitions and also a taxonomy to develop improved robots. He presents the definition of motivation-based robots in opposition to behaviour-based robots, which make use of fixed motivations hardwired in their structure at design time. This kind of robots can sport learning features, but are typically based on a predefined motivation and they learn related behaviours. So the motivation can be to navigate, approach a shelf, move the arm and grab a can. This part must be learned while the motivation is the predefined goal of reaching the can.

On the other side the motivation-based robot does not have predefined motivations provided by design. Motivations are acquired through interaction with the environment. He exemplifies his categorization through real world examples. An automatic camera adjust the parameters to snap a picture according to optimisation strategy. The camera can take thousand o pictures (interaction with the environment) but its internal procedures are not modified by the experience. A toddler exposed to the same environment in which the camera is operating will undergo changes. changes in the memory, in the way he reacts to environment, internal changes modifying the way he will react to future external situations. His motivations are continuously updated and changed.

Midway are some technical solutions like artificial neural networks. For example those exploited for speech recognition. These can change their behaviour in order to improve the discrimination of na individual voice. Nevertheless they act always on the basis of the same motivation (the speech recognition improvement). The proposed taxonomy of robotics architectures has three categories: fixed control architecture, learning architecture, ontogenetic architecture. In the fixed control architecture (the camera example) the system behaviour has been previously coded in the system. What the system is able to do and also in which way the system acts are both predefined.

In a learning architecture (the example of learning artificial neural networks) the system as a fixed goal (or motivation) and this can be considered the phylogenetic part of the architecture (fixed by design or genes) but the way it acts to achieve the goal can be learned, for example reaching a location (the goal) through a cluttered environment or learning affordances of objects through physical interactions like grasping. In the ontogenetic architecture the system learns and evolves in both areas of what goal is relevant (which task to perform) and also in which way (how to) performs the task.

The taxonomy of Manzotti [30] stems form the biological concepts of Phylogeny and Ontogeny, where Phylogeny in relation to behaviour is the hardwired part due to the inherited genes. The ontogeny is the individual development of behaviours generated by the interaction with the environment, by the experience. He implemented an engineering solution to generate motivation in robots with the research objective to test if a robot embodying this architecture would have been able to evolve to new self generated motivations.

The architecture is based on the following modules: The Category module; the Phylogenetic module; the Ontogenetic module. The category module receives external sensory input and categorises it in classes, clusters and categories. It has no restrictions or requirements about the kind of input signals. The phylogenetic module is an equivalent of the genetically predetermined instincts of living animals. It includes predefined criteria for the incoming signals characteristics. It can produce signals because of internal events or because of external events. It includes some functional features prerequisite for certain behaviours (for example move gaze toward more bright objects), it may have one function for each desired behaviour.

The ontogenetic module changes as a result of the interaction with the environment. The criteria contained in this module can be selected on the basis of new experiences. This architecture was exploited in experiments aimed at exploring the possible emergence of new self produced motivations. A robot in which this architecture was implemented, was presented with a set of sundry shapes.

The phylogenetic module as instinctual reaction to coloured shapes. Initially it has no reaction to colourless objects. The experiment was successful in that the robot developed new motivations i.e. it evolved from directing the gaze attention to colourful shapes to a new intentional behaviour directing its gaze attention to specific shapes nevertheless of grey colour (colourless). This motivation was not predesigned by the researcher. So in this way a robotic system can auto-determine. This is one of the most advanced results with regards to autonomously generated evolution of motivation in a real world agent.

Another interesting experience following a similar line of research toward self generated goals and motivations is that of Hanheide [26] about a robot assistant for everyday environments that is capable of doing work for the humans and also in collaboration with them. Their language is less biological-theoretical and more engineering like. They designed their system with the general objective of generating its own behaviour at run-time in opposition to the older way of having behaviour predefined by the programmer at design-time.

This is defined as goal-directed-behaviour, where goal is a concrete state of affairs that the robot shall bring about. They also use the term drive to mean the disposition (i.e. the intention) to achieve the goal. The draw the example of the drive for a waiter to earn as much money as possible that evening, the goals are taking food and drink orders at a certain table, to ask the customers if they are satisfied and so on. In order to be able to pursue the drive the waiter must be capable of Goal management. Their architecture include a Goal Generator module and a Goal management module.

The inbuilt drives are included as Goal generators based on external monitoring of the environment and on the internal states, Goals are produced in order to satisfy to the system drives. Their robot named Dora does an initial tour of the environment to build its knowledge and then in the next round runs in autonomous mode performing actions to implement its goals. Interestingly the system choose to investigate unexplored areas instead of categorise the room in which it ended up, because these areas had an attributed high knowledge gain. The robot system is devised so that the planning activates only one goal at a time. Nevertheless being an interesting architecture, their system falls short of the fact that it does not generates or changes in time its drives. An approach based on advanced mathematical tools is due to Nurzaman [33].

The author work is concerned with the problem of creating a system applicable in robotics that is self-organizing its behaviour. Such includes non deterministic spontaneous dynamics and the challenge is to make this kind of system amenable to the production of desirable behaviours. This approach is called Guided Self Organisation and the background concept is that of homeokinesis that is to generate self organisation modulated by environmental noise or intrinsic noise. In biological system this kind of situation has been studied at the level of sensory systems and also at cellular level.

For robotics system when the approach is that of the embodiment, the behaviour shall merge as the interplay between the physical mechanical structure of the robots, the control system and the environment. The underlying mathematical technique is the so called attractor selection. The robot they use is a very simple curved beam (a vertical metal arch fastened to a horizontal base made out of few beam. The vibrations of the arch due to a rotating mass hooked on a side force the structure to hop. The higher the vibration frequency of the arch, the higher it is the displacement the velocity in the plane) while at the same time the standard deviation of the orientation angle in the plane tends to decrease. At lower frequencies the forward displacement is slower while the robots tends to change direction. So we have two behavioural modes corresponding to two different ranges of frequencies, In such a system the deterministic component can be identified with attractors and the stochastic components are the internal noise and the environmental sensory input. The equations governing the system present two attractors. At low frequency of the motor rotation (that induces the vibrations of the curved beam and the hopping) the speed is low and the direction changes randomly. When the robot is set on his course toward a goal position the presence of bigger noise induces variation in the motor angular velocity, this occurrence shifts the behaviour toward a higher probability of reaching the goal. The noise in the proposed embodiment model corresponds to internally generated noise while another function expresses the sensory feedback input that in case the goal is approaching will tend to persist in the same behaviour.

This contrivance has two basic behaviours that corresponds to two attractors. A change of behaviour is brought about by the noise and causes the system to find another behaviour. The input for the switch is the level of the noise. This model is quite interesting but the results are obtained through simulation. Also it is not defined how the internally generated noise that is involved in the behavioural change would be autonomously varied by a real robot instead of being set ad different levels by the human experimenter.

An approach in someway similar, base on mathematical principles of system exhibiting bifurcations corresponding to different states of the robots, is that of the group of Ralf Der [15, 14]. Their approach was tested on sundry robots. These are embodied and situated, whereas the term situated means that the robot does not loose knowledge of its location and situatedness, or that it does not get lost because of its behaviour and its control based on sensors it is still possible. Their work is based on two fundamental concepts Homeokinesis and Self-organization. The latter is one salient characteristic of natural systems and can be exploited to be build robot which behaviour is self organising. A self organising system is one in which order arises from chaos without external intervention. This approach was already used by Ashby [3].

He developed mathematical descriptions intended to describe a living's organism learning and adaptative intelligence. Homeokinesis is intended as the evolution of the homeostasis concept, i.e. from maintaining a system's internal variables at preset value to one in which the system can evolve through different dynamical behaviours. They Consider a robot with a parameterised controller, a neural network, say, with synaptic weights initially in the tabula rasa condition. So there is no reaction of the robot to its sensor values and activities if present at all are only stochastic ones. The robot is to be in an environment with static and possibly also dynamic objects. The task now is to find an objective which is entirely internal to the robot which drives the parameters so that the robot will start to move and while moving to develop its perception of the world and object related behaviour.

One possibility for the self-organized adaptation is that one uses a supervised learning algorithm for the controller where however the agent is its own supervisor in the sense that he generates its own learning signals Their architecture includes a predictor and a controller. For both of them exist a learning rule.

The drive signal for the learning rule is obtained from the comparison between the predictive model of the behaviour of the agent and the real behaviour in the changing environment. The predictor is concerned with sensory data. It is an internal world model that predicts the sensory data with a certain error. The error is a function that responds to external perturbation. The control is based on a combination of non linearity and of the effect of noise. Noise can be for example the difference between predicted velocity and actual velocity. When the error is small are favoured behaviours. that are well represented by the model. The modelling error is minimized if the behaviour of the robot is unstable in the forward-in-time direction. Inverting the time in the modelling process, i.e. in the prediction of the next step sensors values is the key procedure to create the internal driving force for the self adaptive behaviour generation.

The time loop error is designed as the learning input signal that creates a self-amplification process which in turn is the drive to the generation of new behaviour modes. In this architecture the behaviour depends on initial conditions of the system as well as on environmental conditions. The implementation of this approach on different robots showed the emergent behaviour is due to the interplay of the learning dynamics and of the different embodiments. On the Khepera miniature robot this approach, without any initial preset target behaviour, was able in few minutes to create emergent behaviour in the interaction with the environment, for example pushing a ping pong ball forward, stabilizing it between its front sensors.

Using a Pioneer two wheels robot that includes visual sensory input, when confronted with a suspended ball, the devised two neurons controller was able to start movements autonomously, then to develop visual attention and to also react to the movements of the ball. In the case of a Rocking Stamper robot, a hemisphere with two infrared sensors, a vertical pole and two motors at orthogonal directions acting on it, the experiments showed the emergence of a rocking behaviour of the hemisphere and also a walking-like behaviour. In the case of a five degrees of freedom snake-like segmented robot the emergent behaviour consisted of sits up and occasional jumps. For an hexapod robot, with three degrees of freedom for each leg and an actuator for each of three joint (a total of 18 DoF and servo motors). Proprioception is provided by sensors encoding the joint angle and by torques obtained from a PID controller. The hexapod showed an initial swaying motion and later on other behaviours. such as arising behaviour. Adding exteroception in the form of a sensor to measure forward velocity produced a seesaw behaviour. Most interestingly the learned behaviour made use of inhibitory as well of excitatory signals in connection to the sensors of all the joints of the legs. Also the implementation on a five degree of freedom robotic snake was able to create emergent behaviours.

Kozma [29, 28] claiming a fourth historical period of development in the connectionist approach, widening the classification of Clark [11] implemented a neural layers control model on a Mars Rover obtaining interesting results with regards to intentionality and obstacle avoidance. Essentially their connectionist model is based on hierarchically layered neuronal populations that implement parallel-distributed processing, intended to be an alternative to symbolic systems that oriented to process information serially. They refer to the definition of intentionality of Nunez and Freeman [32]. This is an operational definition stating that it consists of cyclic operation of prediction, testing by action, sensing, perceiving, and assimilation, that becomes apparent because of the dynamical change in the state of the agent through its interaction with the environment.

Their approach is inspired by observations on human electroencephalogram activity human subjects engaged in goal-directed tasks rendered in theoretical form by Freeman [20]. The latter introduced the Katchalsky sets (K sets), a hierarchy of models of the cortical nonlinear neurodynamics related to perception and category learning, They implemented the K-sets model on the SRR2K robot, a mobile platform with four independently steered wheels tested at a Martina soils simulation indoor facility, in order to demonstrate the effectiveness of the reinforcement learning strategy and to demonstrate the feasibility of the proposed method for actual navigation and intentional behaviour. Sensory input is based on measurements by the robot's stereo camera a goal camera mounted on a manipulator arm with 20 degree field of view; an internal DMU gyroscope acting as tilt sensor for pitch, roll, and yaw; The robot was able to learn to avoid obstacles through an exploration activity, to create associations between visual input, vibration (tilt sensor) input and global orientation sensing and to solve problems in perception and navigation that arise by its interaction with the environment as it pursues autonomously the goals selected by its trainer. The implementation included Hebbian correlation learning related to external occasional discrete stimuli acting as reinforcement, reward or punishment, and habituation. i.e. continuous degradation of the response to the environment, unless reinforcing stimuli are intervening.

Baranes [7, 8] introduced an interesting architecture based on the Self-Adaptive Goal Generation-Robust Intelligent Adaptive Curiosity (SAGG-RIAC) algorithm. It has a higher layer of active learning with regards to computation of gaols and self generation of goals, and a lower level of active learning of sensory motor type for computations of more specific Actions Interest and for Goal Directed Exploration and Learning where goals are specific configurations to be reached under constraints such as energy consumption or trajectory of a movement. The higher level of active learning (higher time scale) considers the active self-generation and self-selection of goals, depending on a feedback defined using the level of achievement of previously generated goals. The lower level of active learning (lower time scale) considers the goal-directed active choice and active exploration of lower-level actions to be taken to reach the goals selected at the higher level, and depending on local measures about the evolution of the quality of learnt inverse and/or forward models. They experimented their model with an 8 DOF robotic arm which has goal of exploring the reachable space and learn its forward and inverse kinematics. The arm was provided with low quality motors that have an average noise of 20 % for each movement. The task space was defined as the entire surface visible by the robot camera that is about three times larger than the reachable space. Their algorithm was effective in learning the limits of the reachable area through 10.000 micro-actions and then implement the exploration inside the reachable area, focus toward the ned of the experiment on the most difficult movements to reach areas close to its base. Their approach was also simulated for a quadruped robot and a fishing rod casting the bobber to certain target positions.

Their experiments provided the robot with an initial predefined low-dimensional task space. Their approach in order to be able to support developmental learning and minimize initial setting of parameters by an operator is in need of sundry upgrades.

4.4 CONCLUSIONS

Nevertheless decades long theoretical and practical technological develop-ments and experimenting in cognitive robotics [10, 2, 6, 37, 31, 22, 21, 9], the embodiment approach as a solution to autonomy, self organisation, goals generation, is still limited to some basic features mainly related to lower level sensory motor tasks goals and learning. The recent attention to emotion and motivation as means to improve robotic intelligence and performance accord-ing to the paradigm of human agency provide some interesting initial basic results such as those of Manzotti.

The rationale for their use is nicely and clearly expressed by Arkin [2]: "We instead ask the question, what capabilities can emotions, however defined, endow a robot with that an unemotional robot cannot possess?", "Unfortu-nately, emotions constitute a subset of motivations that provide support for an agents survival in a complex world. They are not related to the formulation of abstract goals that are produced as a result of deliberation. Motivations and emotions affect behavioural performance, but motivation can additionally lead to the formulation of concrete goal-achieving behaviour, at least in hu-mans, whereas emotions are concerned with modulating existing behaviours in support of current activity. In this regard, motivations might additionally invoke specific behaviours to accomplish more deliberative tasks or plans (e.g., strategies for obtaining food)".

In robotics these possible contributions of emotions are yet to be explored in order to fully understand if they are really needed and useful. All the cog-nitive embodiment, emotion and motivation developments in robotics are aimed to improve the intelligence and autonomy that indeed means more adaptability and especially discretionality in behaviours elicited by a chang-ing and challenging environment. The almost unasked and absolutely unan-swered question from the experimental point of view is if the implementation of emotions, motivations and intentional states in robots will produced robots which intelligence and performance will be improved or if they will resent the limits of performance proper of the human being. Also remains an open research question if the increased capacity of goals generation and autonomy will produce behaviours that are amenable to humans expectations and hu-man profitable use of the robots or if we will get robots that will be to a certain degree cleverer but less predictable.

As lively expressed by Haikonen [34]: "Machine emotions would involve instant judgement by emotional value, system reactions and direction of action as well as motivational effects. In principle these would be useful func-tions but sometimes might counteract more appropriate rational responses. Quick emotional short-cut reactions in a dangerous situation may save the day for the robot, but a robot in emotional rage would be no good for most purposes. It is up to the designer to find a proper balance here." In the case of self learning robots, if the pattern of behaviour acquisition were deter-ministically determined possibly we would not have the individuality of the choice (two individual robots would choose the same behaviour or develop the same habit or way of life in relation to a wide set of possible alternatives (this is like if all the children in an ice cream shop will choose the same set of flavours and whenever entering an ice cream shop they all will choose again the same).

So to have intentionality, individual choices, we will need to have differences between robots interactions with the environment and with the humans, and consequently different embodiments and different behaviours and so far an element of alea. At what extent will the approach to self learning, self determined, self organising agents realize unpredictable (at least within the allotted boundaries of alternative choices compatible with survival of the robot and with environment sharing without damage to humans, animals and objects and goods) behaviours?

4.5   Bibliography

[1] Michael L Anderson. Embodied cognition: a field guide. *Computing Reviews*, 45(5):304, 2004.

[2] Ronald C Arkin. *Moving up the food chain: Motivation and Emotion in behavior-based robots*. Georgia Institute of Technology, 2003.

[3] W. R. Ashby. Principles of the self-organizing dynamic system. *Journal of General Psychology*, 37:125–128, 1947.

[4] W. R. Ashby. *Design for a brain*. Wiley, Oxfors, England, 1952.

[5] Murat Aydede and Philip Robbins. *The Cambridge handbook of situated cognition*. Cambridge University Press New York, NY, 2009.

[6] Gianluca Baldassarre, Tom Stafford, Marco Mirolli, Peter Redgrave, Richard M Ryan, and Andrew Barto. Intrinsic motivations and open-ended development in animals, humans, and robots: an overview. *Frontiers in psychology*, 5, 2014.

[7] Adrien Baranes and Pierre-Yves Oudeyer. Active learning of inverse models with intrinsically motivated goal exploration in robots. *Robotics and Autonomous Systems*, 61(1):49–73, 2013.

[8] Oudeyer P. Y. Baranes, A. Intrinsically motivated goal exploration for active motor learning in robots: A case study. In *Intelligent Robots and Systems (IROS), 2010 IEEE/RSJ International Conference*, pages 1766–1773. IEEE, 2010.

[9] Justin Blount, Michael Gelfond, and Marcello Balduccini. Towards a theory of intentional agents. In *Knowledge Representation and Reasoning in Robotics, AAAI Spring Symp. Series*, 2014.

[10] Donald S Borrett, David Shih, Michael Tomko, Sarah Borrett, and Hon C Kwan. Hegelian phenomenology and robotics. *International Journal of Machine Consciousness*, 3(01):219–235, 2011.

[11] A. Clark. *Mindware: An Introduction to the Philosophy of Cognitive Science*. Oxford University Press, 2001.

[12] Silvia Coradeschi, Amy Loutfi, and Britta Wrede. A short review of symbol grounding in robotic and intelligent systems. *KI-Künstliche Intelligenz*, 27(2):129–136, 2013.

[13] A. Damasio. *The Feeling of What Happens: Body and Emotion in the Making of Consciousness*. Harvest Books, New York, USA, 2000.

[14] Ralf Der, Frank Hesse, René Liebscher, et al. Self-organized exploration and automatic sensor integration from the homeokinetic principle. In *Proceedings of Workshop on SOAVE*, volume 4, pages 220–230, 2004.

[15] Ralf Der and Georg Martius. Behavior as broken symmetry in embodied self-organizing robots. In *Advances in Artificial Life, ECAL*, volume 12, pages 601–608, 2013.

[16] E. A. DiPaolo. Robotics inspired in the organism. *Intellectica*, 53(1):129–162, 2010.

[17] H.L. Dreyfus. *What Computers Cannot Do: A Critique of Artificial Intelligence*. Harper and Row, New York, 1972.

[18] P. Ekman. An argument for basic emotions. *Cognition & Emotion*, 6(3-4):169–200, 1992.

[19] E. Rosch F. Varela, E. Thompson. *The Embodied Mind*. MIT Press, Cambridge, MA, 1991.

[20] W.J. Freeman. *Mass Action in the Nervous System*. Academic Press, New York, 1975.

[21] O. L. Georgeon. Learning by Experiencing versus Learning by Registering. *Constructivist Foundations*, 9(2):211–213, 2014.

[22] Olivier L Georgeon, James B Marshall, and Riccardo Manzotti. Eca: An enactivist cognitive architecture based on sensorimotor modeling. *Biologically Inspired Cognitive Architectures*, 6:46–57, 2013.

[23] J. J. Gibson. *The ecological approach to visual perception*. John Wiley & Sons, Inc. New York, NY, USA, 1979.

[24] K. Goldstein. *Der Aufbau des Organismus. Einfuehrung in die Biologie unter besonderer Beruecksichtigung der Erfahrungen am kranken Menschen*. Nijhoff, Den Haag, 1934.

[25] Paul E Griffiths. *What emotions really are: The problem of psychological categories*. Chicago: University of Chicago Press, 1997.

[26] Marc Hanheide, Nick Hawes, Jeremy Wyatt, Moritz Göbelbecker, Michael Brenner, Kristoffer Sjöö, Alper Aydemir, Patric Jensfelt, Hendrik Zender, and Geert-Jan Kruijff. A framework for goal generation and management. In *Proceedings of the AAAI workshop on goal-directed autonomy*, 2010.

[27] et al. H.Choxi. Using motivations for interactive robot behaviour control. *DARPA contract 66001*, 01(D):169–200, 2006.

[28] Hunstberger T. Aghazarian H. & Freeman W. J. Kozma, R. Implementing intentional robotics principles using SSR2K platform. In *IROS 2007 IEEE/RSJ International Conference*, pages 2262–2267. IEEE, 2007.

[29] Robert Kozma, Hrand Aghazarian, Terry Huntsberger, Eddie Tunstel, and Walter J Freeman. Computational aspects of cognition and consciousness in intelligent devices. *Computational Intelligence Magazine, IEEE*, 2(3):53–64, 2007.

[30] Riccardo Manzotti and Vincenzo Tagliasco. From behaviour-based robots to motivation-based robots. *Robotics and Autonomous Systems*, 51(2):175–190, 2005.

[31] V. C. Mueller. Autonomous cognitive systems in real-world environments: Less control, more flexibility and better interaction. *Cognitive Computation*, 4(3):212–215, 2012.

[32] Freeman W.J. Nunez, R.E. Restoring to cognition the forgotten primacy of action, intention, and emotion. *J. Consciousness Studies*, 6(11-12):ix–xx, 1999.

[33] Surya G Nurzaman, Xiaoxiang Yu, Yongjae Kim, and Fumiya Iida. Guided self-organization in a dynamic embodied system based on attractor selection mechanism. *Entropy*, 16(5):2592–2610, 2014.

[34] Haikonen P. O. *Artificial minds and conscious machines*. Information Science Publishing USA, 2005.

[35] J. Piaget. *Biologie et Connaissance: Essai sur les relations entre les regulations organiques et les processus cognitifs*. Gallimard, 1967.

[36] Tiago Ribeiro and Ana Paiva. The illusion of robotic life: principles and practices of animation for robots. In *Proceedings of the seventh annual ACM/IEEE international conference on Human-Robot Interaction*, pages 383–390. ACM, 2012.

[37] Alexander Riegler, Markus F Peschl, and Astrid von Stein. *Understanding representation in the cognitive sciences*. 1999.

[38] John Sabini and Maury Silver. Ekman's basic emotions: Why not love and jealousy? *Cognition & Emotion*, 19(5):693–712, 2005.

[39] J. R. Searle. Minds, brains, and programs. *Behavioral and brain sciences*, 3(3):417–424, 1980.

[40] F. J. Varela. *Principles of biological autonomy*. Elsevier, North Holland, New York, 1979.

[41] Dave Ward and Mog Stapleton. *Consciousness in interaction*, volume 86 of *Advances in Consciousness Research*, chapter Es are good. Cognition as enacted, embodied, embedded, affective and extended, pages 89–104. John Benjamins Publishing Company, Amsterdam/Philadelphia, 2012.

[42] T. Ziemke. Are robots embodied. *In First international workshop on epigenetic robotics Modeling Cognitive Development in Robotic Systems*, 85:701–746, 2001, September.

CHAPTER 5

Imagined Physics: Exploring Examples of Shape-Changing Interfaces

Tim R. Merritt

Aarhus School of Architecture, Denmark.

Mie Nørgaard

MIE NØRGAARD, Denmark.

Christian Ø. Laursen

Aarhus University, Denmark.

Majken K. Rasmussen

Aarhus University, Denmark.

Marianne G. Petersen

Aarhus University, Denmark.

CONTENTS

I n the field of cognitive robotics, much attention has focused on the technical challenges and approaches to building social robots and intelligent agents. However, few works explore how the aesthetic qualities of shape change influence how people perceive such an agent. Our contribution to this book focuses on the human responses to objects that change shape in response to input from users, environment, or other circumstances. In this chapter we discuss the term *imagined physics*[1], meaning how actuated devices are in one sense tied to their physical form, yet through the use of actuators, sensors, and computer algorithms can behave in ways that are surprising, unpredictable and that might even be perceived as magical. We also claim that shape change can help reveal the state of a robot or object, providing cues for mentalizing, similar to how we read emotions and understand body posture and other non-verbal communication when interacting with people. We review examples of shape-changing interfaces including toys, interactive lights, robots, etc, noting the intentions – or claims about intent - made by the respective designers.

5.1   INTRODUCTION

Across examples of research in the field of cognitive robotics, researchers embrace goals of designing robotic systems that can perceive and react to humans and deal with complex situations. We turn that around and propose that there is also important research needed in designing robots such that the *robot's* intentions, goals, and even *mental states* [26] can be more effectively inferred or understood by humans. This is important for situations in which humans need to understand and cooperate with a robot toward a common goal. Few works explore how the aesthetic qualities of shape change influence how people perceive autonomous agents. This work examines some of the building blocks of robotics–objects that move and components that make shape-changing interfaces possible. We make a small step in that direction and in this chapter, we share and discuss the design of shape-changing interfaces and how experiential qualities can influence the interaction [28].

The rest of this chapter is organized as follows: firstly, we discuss shape-changing interfaces as a research topic, then we provide examples from our design courses focused on shape-changing interfaces, we then describe the concept of "imagined physics", which designers can use to accentuate, exaggerate, or attempt to hide behaviors, current state, and intentions of complex autonomous robots. We then provide a short analysis of typical electronic components that have been used in the various projects designed by

---

[1]This chapter is based in part on our paper [28].

participants in our courses noting key characteristics of the movement supported and highlight the incidental effects that these technical elements introduce (sound, smoothness, heat, etc). Additionally, we describe the design challenges from our experiences in dealing with these contingencies and leveraging them in the design choices. We briefly propose future research to continue the exploration in the field of shape-changing interfaces.

5.2 SHAPE-CHANGING INTERFACES

The term, "shape-changing interface" involves *self-actuated change*, which "must be controllable so that the object can return to its initial state and repeat the shape change" as defined in [40]. In that paper, the authors note various titles used to describe interactive artifacts that change shape including terms such as *actuated interfaces [37], kinetic interaction [33], organic user interfaces [7], kinetic organic interfaces [32], pro-active architecture [29] and computational composites [46]*.

There are many examples of shape-changing interfaces in recent research and with the continued advancements in materials and electronics, this is likely to continue. Many examples in the field are discussed in [40], and since then, a rich body of work in the field includes systems of various scales including hand scale of mobile phones [35], aesthetic experience pieces that encourage interaction and play [3], shape changing seating to influence atmospheres in public spaces [13], and experiments in the architectural scale with shape-changing rooms [45], among others.

Shape change holds potential beyond the scope of playful objects, from small scale applications, such as dynamic buttons [14], shape-changing mobile phones [15, 16], or wiggling attention seeking post it's [38] to large scale dynamic architectural elements, kinetic facades [24], shape-changing architectural structures [29], or visions of permeable architecture [6]. These are just fragments of the abundance of examples, which have sprouted from the field within the last decade.

Beyond the multitude of point designs, exploring the potential application areas, experiences and interaction that making the physical form dynamic enables, some papers have also pointed to the challenges, potential and limitations of the current technologies [9], providing toolkits [31], framing different aspects of kinetic vocabularies [33, 40], introducing new technologies [10] and pointing to future visions for the field [17].

5.3 EXAMPLES FROM OUR COURSES

In this chapter we examine projects that researchers and students taking part in our design classes have created. This allows us to, more closely, examine the challenges the designers faced and provides insights intimate to the design process. Examples include the outcomes from two courses focused directly on designing shape-changing objects including tumbling objects in one course, light in another. We also include additional projects from an innovation course and independent study project relating to shape-changing surfaces.

In the following subsections, we account for the examples, how they work, what purposes they fulfill and which experiential qualities they provide. The examples have been realised through functional prototypes, in which users can experience these shape-changing interfaces first-hand.

5.3.1 Shape-Changing Tumbling Objects: bObles

There are many example of HCI research featuring the integration of sensors, actuators and computing power into playful objects for children. From animated plush toys (e.g. [43]), to digital playgrounds (e.g. [41, 42]), animated paper [21], shape-changing toys (e.g. [19, 39]), open-ended interactive play objects (e.g. [4, 44]) and many other inspiring playful interactive objects and concepts. Topobo [[39] and Kinematics [30] present playful toolkits allowing users to explore movement and physical change through passive and augmented building blocks. Ninja Track [19] presents a different approach to shape-changing toys, whereas the shape change is used to switch between two states of play. It can change from a flexible whip into a rigid sword with the push of a button. There are other examples of augmented objects that use shape change to facilitate playful interactions, for example the physical kinetic surface of Kinetic Bricks [20] could be used as a construction toy for children, and SpeakCup [48], which functionality could be used as a toy for social interaction in public spaces inviting the user to speak and listen according to the shape they have formed with the object.

Among the various examples of shape-changing toys, we conducted a design course with the support of the Danish design company bObles in order to explore how the various types of shape change could support playful interactions. bObles designs playful tumbling objects for children, following a design philosophy that dictates simplicity in material and concept, and aim to inspire imagination and movement with their products. The tumbling objects are made from solid EVA foam blocks and are quite versatile; they work equally well as furniture, building blocks and obstacles in a game of tag, or whatever a child might imagine.

The course participants were assigned the task of redesigning a bObles tumbling object using one specific type of shape change. Additionally, the design should meet with bObles design philosophy. The shape change framework [40] was used as a way to open the design space, and broaden the exploration of shape-changing interfaces. Each group received one of the eight types of shape change as a constraint and were told to involve that type of shape change in some way. Participants were urged to develop their understanding of the properties of shape change through (hardware) sketching, drawing directly on Buxton's [5] understanding of the praxis. The making of sketches was supervised and critiqued on an ongoing basis by a team of interaction designers challenging design choices and the qualities derived from these. Materials in addition to the foam objects included an array of actuators (linear, servos, stepper motors, lights, fans, bubble machine, vibrotactile actuators, etc.), sensors (pressure, light, proximity, sound, etc.), which could be controlled using Arduino microprocessor boards.

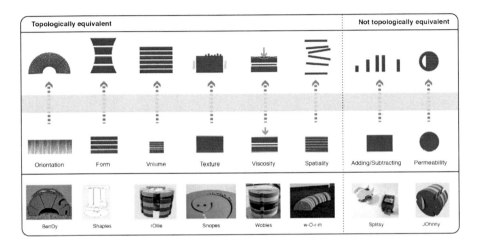

FIGURE 5.1 Overview of eight shape change sketches.

Students developed a wide range of physical sketches that illustrated different ways that shape change could be initiated by the human user often involving metaphors from the physical world. One such example, "rOllie," used the metaphor of a snowball to map user behavior with the volume change events. rOllie increases in volume when rolled on the floor, mimicking how a snowball gathers snow and becomes larger as shown in 5.2. To decrease volume, the user pats the surface as if knocking off the snow and returning the object to its original shape. A microphone placed inside the cylinder registers different patterns of patting, making it possible to shrink the volume according to, for example, how hard the cylinder is being hit or with what frequency.

The designers encountered various challenges and opportunities presented by the actuators and incidental effects of controlling the movement in their respective projects. For example, in the sketch called, "jOhnny," the designers worked with the changing sound of the servos under changes in the load. The slow yet labored movement of the gills of the fish-like robot was supported by the slight changes in pitch when switching between opening and closing, resulting in a cohesive impression that the fish was struggling

FIGURE 5.2 rOllie toy concept illustrating metaphor of rolling to add volume and patting to reduce volume.

and labored. While this sound was used in a supportive way to the design intention, in other cases, designers struggled to minimize actuator noises. Only through building and embedding the components during the design process can these issues be noticed and dealt with by the designers.

Following this first course we proposed a design space called imagined physics[28], in which the design can be described by how a user would perceive the shape change. This model, which is described in more detail later in this chapter, classifies shape-changing elements according to what is made visible or hidden, what is virtual or physical, and how volatile or consistent the perceived laws of physics are which govern the movements. We designed the course for the subsequent year to explore this concept further.

5.3.2 Shape-Changing with Light

The design course set out to explore light and shape-changing interfaces and to explore the concept of 'imagined physics', in order to create a shape-changing interface that could alter its shape according to some input or data. The student groups were given the freedom to choose any of the eight types of shape change in their design work and were encouraged to push the boundaries from a research perspective by considering the call for submissions to the TEI 2014 art gallery [1], which proposed submissions that question, "how can we incorporate the tangible, embodied and embedded technologies that bring awareness to ourselves as individuals and as communities?" Two of the groups submitted and were accepted to the TEI art gallery of 2014 to display their work (including one author of this chapter).

Kinetic Wave [36] is an interactive, shape-changing installation that explores how input from its' surroundings can be used to create abstract reactions that communicates information through the use of spatial shape change. Kinetic Wave aims to visualize the invisible wireless communication caused by everyday, wireless devices such as smart-phones, routers and laptops - a dimension not perceivable by the human sensory system.

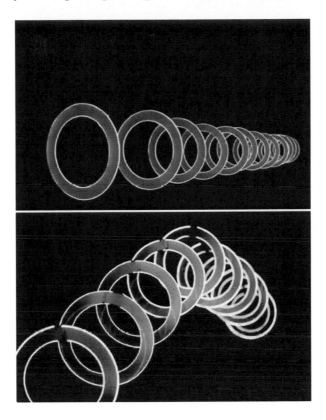

FIGURE 5.3 The illuminated rings resembling a wave, which changes amplitude based on ambient electromagnetic signals.

The installation consists of 12 frosted, acrylic rings with electroluminescent wire on the inner and outer edges (see Figure 5.3). The rings are suspended from a shared platform, where each ring is attached to a stepper motor that controls the vertical movement. The combined movement of the rings resembles a wave that creates an abstract visualization of invisible radio waves. This gives the audience an illusion of the invisible radio waves propagating into the physical, visible domain of the installation, where the amplitude of the symbolic waves is dependent of ambient wireless activity.

For the user, it seems as if the laws of physics does not apply to the installation. The mechanics are hidden and the audience are left to imagine how the wireless communication affects the installation as a whole. The experiential qualities of the installation are affected by having the mechanics (stepper motors and so forth) hidden from the users, which is complimented by the mystifying sound of the stepper motors accelerating/decelerating - which allows the user to imagine the mechanics behind. It is especially seen how the lights of the installation leaves a lot to the imagination of the beholder, as it is not easily seen how the light is diffused into the acrylic rings, where the light source comes from, nor is it quickly seen where the light is powered from - as the installation is exhibited in a dark setting.

The *Sensitive RolyPoly* [12] is another work that was selected for the gallery from the course. The designers proposed that the shape change activity would connect to people by visually displaying ambient sound levels.

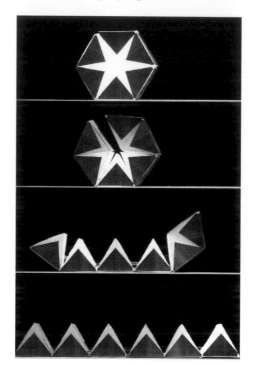

FIGURE 5.4 The different states and transformation of the RolyPoly, (from top) with high levels of ambient noise, the object curls up, yet as the ambient noise decreases, it changes shape to a straight form (bottom).

The RolyPoly can curl up, forming a hexagon or extending itself into a line of triangles. The transformation appeals to the viewer through zoomorphic characteristics. However, contrary to natural characteristics, the behaviour of the RolyPoly is reversed, such that if the environment has a high level of ambient noise, it will fully extend, making the triangles spike upwards in a defensive manner (See Figure 5.4). As a way to grab attention, the light at each tooth flickers. If the ambient noise level is low, the RolyPoly will curl up—at the same time, the internal lights begin to pulsate, drawing from a 'breathing' metaphor.

The RolyPoly functions as a ambient display, or reminder, of the noise in its surroundings, by using zoomorphic behaviour to communicate either chaos, through the sharp edges of the teeth, to the more pleasing hexagon from which diffused light fades in and out. As opposed to the Kinetic Wave example, the RolyPoly makes use of the expressive parameters of shape-change through a 'being' as the focal object, which exhibits traits intended for the human to relate to emotions, such as anger and happiness.

5.3.3   Shape-Changing Surfaces

In addition to the larger design courses, specialised study groups have focused on examining how users perceive the kinetic parameters of shape-changing interfaces. We will briefly describe an example of this work in which the students designed and built functional prototypes to support user evaluations of interactions involving pixel-based and homogeneous shape-changing surfaces.

Two master students have researched into how users perceive the interaction, motion and utility of shape-changing interfaces and evaluated prototypes through simple user evaluations. [2]

The study was conducted with functional prototypes surfaces that supported and actuated a mobile phone that was placed on it. One system is pixel-based, with a control console comprised of switches, sliders, and dials that enable the study participant to make adjustments to the sensitivity of the interaction, speed, and modes of actuation. The other prototype system presented a homogenous surface with a membrane covering that hides the actuation mechanics to facilitate a comparison for the participants.

The pixel-based prototype consisted of 100 tiles, in which 32 were actuated, the remaining 68 did not move, yet served as visual cues to suggest a high resolution interaction area. The 32 tiles were placed in an 8 x 4 grid, where each row of 4 tiles were interconnected. Each row could actuate 4 cm vertically below the surface of the prototype and 5 cm above, a total of 9 cm linear actuation.

The evaluations were conducted in a public setting, whereas passers-by were invited to participate in a pre-defined use case with the pixel-based prototype. This served as a way to test how users perceived and understood the shape-changing interface by interacting with a smartphone and the surface as an inductive charging platform as the setting for the experiment. The prototype served as a vision of how a smartphone could be placed on a surface for charging, whereas the shape-changing surface would actuate according to the battery level of the smartphone. If the smartphone was low on battery

---

[2]This work was conducted by Stephan Sloth Kristiansen and Kennet Jeppesen under the guidance of Marianne Graves Petersen and Tim Merritt from 2013-2014.

power, it would sink beneath the baseline of the table and actuate upwards as it slowly charged until it became fully-charged and aligned with the surface (see figure 5.5 - middle), allowing the user to easily pick it up. The experiment served to answer three questions about users and shape-changing surfaces. Firstly, to examine user reactions to interacting with this type of interface. Secondly, to examine the users' understanding of the surface actuation in regards to parameters of transformation (speed, non-repetitive/repetitive) and the aesthetic qualities of these. Thirdly, to investigate how designers could design for meaningful correlations between input and output, and meaningful interaction between the users and shape-changing surfaces.

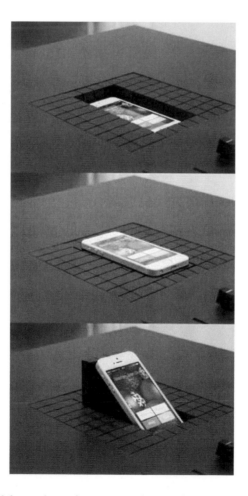

FIGURE 5.5 Pixel-based surface uses shape change to support interactions with the mobile phone.

The pixel-based surface could be accessed by reaching for it, which triggered the vertical actuation, raising the smart-phone towards the palm of the users (see Figure 5.5 — bottom).

Preliminary findings showed that the presence of hand as an interaction modality suited the interface of a shape-changing surface as objects are typically placed or taken from surfaces. The initial impressions of the prototype were concentrated on how the relation between the movement of the user and that of the interface affected the experiential qualities of the system as a whole. The users' perceptions of the shape-changing surface is dependent on the use-case, the velocity and path parameters, such as speed, acceleration and pattern of the transformation.

There were insights gathered relating to the emotions that participants perceived according to the speed of the transformations–fast transformations can be seen as an aggressive reaction towards the user, while a slow transformation could lead to the diminish belief in the perceived capabilities of the surface. However, this could also be experienced as a calm reaction - again entirely dependent on the context. There were many participants who claimed that the speed should in some way match the speed of the reaching hand in order to feel appropriate.

The homogeneous prototype was built with two linear actuators mounted at each end of a MDF plate, using hinges to support sloped surfaces. The plate was covered with flexible spandex fabric stretched over the top, allowing it to tilt back and forth as well as moving vertically. Due to limitations in the spandex flexibility, it can actuate roughly 15 mm below and above the baseline of the surface. When the MDF plate was aligned with the baseline of the surface, it could be perceived as a continuous surface. As the mechanics are more hidden in contrast to the pixel-based interface, it becomes more challenging for the user to understand the mechanics behind the movement, which allows for the user to be more imaginative about how it works and the limitations of the interface.

The pixel-based prototype used servos to actuate each row, which caused incidental sound effects that influenced the perception of the movement. Often participants imagined a link between the movement and the sound generated, remarking on the noise that seemed to be related to the speed of actuation–louder whines with the fast movement were perceived as aggressive and mechanical. In the pixel-based prototype, inexpensive servos with plastic gearing were used, thus a high-pitched sound is emitted from the DC motor and the increasing friction caused by the high speed movement of the gears.

5.4   IMAGINED PHYSICS

The work presented above suggests how the sketches are in one sense tied to their physical form, yet through the use of actuators, sensors, and computer

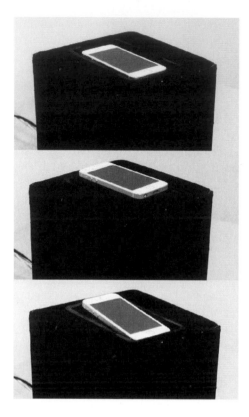

FIGURE 5.6 The homogeneous shape-changing surface prototype when actuated below the surface (top), actuated above the surface (middle) and actuated to a inclined position (bottom).

algorithms can behave in ways that are surprising, unpredictable and that might even be perceived as magical. The concept sketches enable the designers to try different behaviour mappings, input modes and transitions, and go beyond traditional physics to an imaginary world of other rules. We call this phenomenon 'imagined physics' in want for better words.

Adding imagined physics to real physics presumably gives rise to new experiences that are initially tied to metaphors based on real physics. For example, rolling rOllie results in volume growth just as when rolling a snowball, whereas the patting used to reduce the rOllie's size, does not draw on either familiar physics or hold a connection to the physical properties of snow and snowballs. However, it is quite similar to the physical and virtual elements discussed in [2] with the physical toy duck that follows the user's hand keeping a consistent distance thus leading the user to imagine an invisible leash. In the case of the rOllie sketch, the interaction does not entirely build on our experience from real world physics, the patting to reduce size-interaction seems plausible and easily understood, however the user must fill in what is not made visible in the materials.

The imagined physics can borrow from metaphors in shape and nature. In the zoomorphic examples the child is invited to learn the "personality" of the object, for example, the worm utilized sensors that look like eyes, and the behaviour drawing away from the approaching hand gives the impression of managing its personal space. This design's connection between the cause and effect is rather simple, and is suggested through cues found in living things. When we move away from these cues and instead map user input/actions to shape change behaviours we find examples that could be explained through imagined physical phenomenon. *Shuples*, for example, changes between two form states depending on the roughness of handling. Like with the Etch a Sketch drawing toy that hides the sand that enable erasing the drawing, a child will quickly learn to interact with the product though the this mapping is not explicitly signalled. For some of the sketches, real physics are manipulated and managed in ways not possible to mimic with physical objects. For example, *Wobles* involved the "programmed" viscosity of the object, in this case mapping the roughness of handling to firmness, however since Wobles does not make use or 'real' physics, that connection could have been tied to any input.

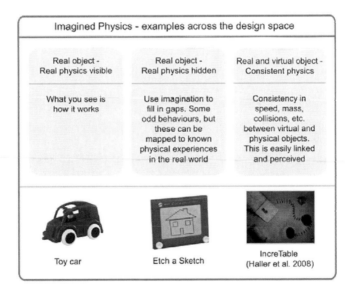

FIGURE 5.7   Three examples across the *Imagined Physics* design space.

As a preliminary step in explaining this concept, we provide three examples with distinct differences shown in Figure 5.7. The first example, the toy car, contains parts that are entirely physical with movements governed by physics that are visible, consistent, and easy to understand. We can easily imagine that the wheels is fixed to the body of the car and rotate around the axles. However, when we consider the *Etch a Sketch*, it involves only physical elements, yet the mechanics that connect the turning of the knobs to the drawing that appears on the screen is hidden. Through use, the operator develops a mental model for mapping the movement to the drawing that unfolds, the left knob always moves the cursor vertically and the right knob always moves the cursor horizontally.

In the example of the IncreTables [22] system, there is a mix of physical and virtual objects that build the experience. In that system, the table surface is a display screen that is augmented with sensing to connect the virtual and real world events. The table surface appears as a virtual landscape with virtual dominoes. On top of the table, physical dominoes are arranged in patterns that join up to the virtual dominoes. The system supports the crossing over of the dominos falling from the virtual and real world–thus a virtual domino can knock over a real domino in the game experience. In that system, the designers chose to match the physics that govern movements to seem as consistent as possible in the virtual world and physical world. However, since the movements are not intrinsically connected to the laws of physics, the designers could introduce behaviors that challenge the expectations of the participant.

The speed of the virtual objects could be adjusted, going beyond what is possible with phyiscal objects, collisions could be randomized or behaviors of the dominos falling could become more volatile in various ways. It is possible that with virtual elements shape-changing interfaces can involve designed elements that are unpredictable, volatile, and include behaviors that seem to change without causality and that would require intense imagination if not to cause complete confusion.

In taking a step back, it seems that the range of objects could be placed in a three dimensional design space as shown in 5.8 with axes describing the material and interactive features physical/virtual, volatile/consistent, and visible/hidden. Objects could involve physical or virtual elements or a combination as in IncreTables [22]. The rules of physics could be consistent and predictable or volatile (changing erratically). The third dimension deals with causality, the rules of physics can be shown to the user and made visible through form elements as is the case with a bicycle chain and sprocket, or they can be hidden as is the case with "black box" technologies that hide the mechanics of physical behaviors.

More in depth analysis of interactive systems according to these diagrams will likely lead to refinements of the models - we hope this future work will also inform the design of inspiring shape-changing objects. For example, it may be difficult to consider an object that is mainly physical, with visible mechanics enabling the shape change yet that are volatile in their behaviors. The sketch Splitsy, if made according to the designers' vision of using programmable matter, could be such a system.

The concept of imagined physics is interesting because it treads on HCI

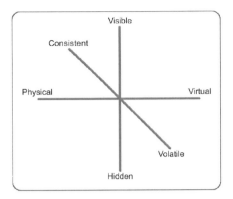

FIGURE 5.8 Design space of Imagined physics as three dimensional space.

territory that may conflict or seem at odds with expected form language and interface affordances. With objects embedded with "imagined physics" it is unclear how much the user will continue to explore and try to find new connections between their behaviour and the shape-changing object's behaviour unless it is suggested in some way. In game design, for example, the length of the experience is suggested as an estimated play time on the casing, but in games clear goals are often achieved along the way to signal the progression, which help a user stay engaged in the exploration of the game. In hypertext fiction research, studies suggest that users explore for some time, but quickly change their focus toward "getting the gist" of the story [23]. In those studies, users were faced with obstacles and unexpected behaviours of the storyworld controlled by the computer system. At first, users would explore and look for variation; however, there was a rather quick shift toward making sense of the story that all users shared. Given playful shape-changing objects that can change shape according to mappings that may change and may not adhere to 'real' physics, it would be interesting future work to evaluate whether users want continued and surprising imagined physics, or do they strive to develop an understanding of the general behaviour of the object and if this understanding of how things work is upset by changes, how tolerant are the users or does it draw them in even more?

5.5 MOVEMENT AND INCIDENTAL EFFECTS OF ACTUATORS

Robots and shape-changing objects take action in the world through a variety of electronic and mechanical means. Skillful designers choose the actuators carefully according to the physical characteristics and performance taking into consideration the type of movement desired (rotation, linear, etc.). Perhaps the most common constraints include actuator size, weight, mounting requirements, efficiency, power consumption, material composition, control characteristics, among others. There continues to be advances in research showing improvements to the fluidity of movement and reducing actuator noise [47], however, for the purposes of this chapter, we draw focus on the movement and other aesthetic qualities involved with existing and commonly available actuators. Through observation and feedback from the designers in our courses, we gathered details about some of these incidental effects involved with typical actuators including considerations of sound, heat, and qualities of movement. This contribution is not comprehensive, but we hope that it fosters discussion about choosing actuators to support design intentions and to consider the incidental effects as a resource to be embraced. We discussed this in [28], however we provide more technical insights here and invite designers to consider not only their choices for actuators, but to look for ways to design the use of the actuators exploiting the underlying potentials inherent. For example, industrial designers focusing on psychoacoustics have examined the perception of sound associated with car doors, electric train motor noise and how this can elicit emotions and perceptions [8, 34].

This table is very brief and distills some of the challenges faced by the designers in the work presented earlier in this chapter and from other projects in our classes. It can be useful for designers to scan when selecting actuators, to identify what are likely challenges with a chosen technology, and when a desired effect is in mind–helping to find actuators that might support these needs through the inherent qualities. We now briefly discuss an example of the design issues related to a project which helped to build this table.

The Kinetic Wave project utilized stepper motors to raise and lower individual rings by winding and unwinding wire from spools. The designers chose the stepper motor for its ability to support smooth and precise movements that could be coordinated among the 12 rings, however, they took much care in dealing with the sound of the motor movements. At certain frequencies, and as the speed increases, there are 'bands' in which resonant frequencies are reached resulting in amplified noises that would cause a buzzing sound in the frame of the art piece. The designers then used rubber bushings to mitigate the effects of the noise, however, they took care in crafting the control curve for moving the rings to minimize the noise. They did however experiment with controlling the stepper motors to strike a balance between speed and resulting sound, thus accentuating the light and movement with the sound characteristics. Other challenges that were faced in the project involved the linkages and joints that connected the stepper motors to the spools. The designers used many laser cut parts that, in principle, should be very accurate, however, the cuts made in different materials (wood, acrylic, MDF) resulted in different amounts of material consumed by the cut. Although the differences were small, these small differences introduced a noticeable wobble that would produce a secondary movement when rotating

Actuator	Movement Type	Response time	Smoothness	Sound	Integration issues
DC servo motor	rotation	fast	quality of components	gear noise	linkage, heat
SMA servo	rotation	moderate	very smooth	silent	high current, slow
SMA wire	contraction / trained shape	fast activation, slow return	smooth	silent	heat, high current, slow
stepper motor	rotation	fast	smooth at low speeds	humming or harmonic resonance	linkage, alignment, speed, jitter
linear actuator	linear	fast	smooth	soft motor whine	linkage, speed, load
solenoid	linear, rotation	fast	instant movement hard stop	click	heat, high current, abrupt
vacuum (jamming)	fluid	less than a few seconds	smooth	pump noise	equipment size
air pressure	fluid	less than a few seconds	smooth	pump noise	equipment size
fan	fluid	moderate	smooth	wind noise	control of air

TABLE 5.1   Design considerations for typical actuators.

the spools. The designers had to work carefully to minimize this through additional iterations of the spool design and cutting process.

5.6   FUTURE WORK

We have briefly reviewed a number of examples of shape-changing interfaces noting aesthetic qualities that can impact the perception of the object and experience of the interaction. There are ever more complex and advanced materials being developed and computational power is becoming embedded into the materials around us [17]. Designers are searching for ways that the formal characteristics of the objects they bring into the world connect with people emotionally and perceptually from explorations of analog materials [27] to interactions in smart environments [25].

It is important that we consider and explore metaphors that can be effective and desirable for people to share meaning with robots. An obvious next step is to conduct research into the user experience through actual use of shape-changing artifacts to help guide future refinements. There is much needed work, which we envision can include a mix of methods. There should be continued exploration following a research through design approach [49, 11], identifying new contexts that might be helpful for shape change, framing the problems and opportunities to be addressed, exploring materials, and sharing the results from the process of making so that others might learn from the challenges and take inspiration from detailed accounts of the designers' leaps of creativity. There are many recent examples of shape change research from mobile phones to objects of larger scale and environments, often with short-term evaluations or initial impressions. It is also important that design researchers embed these technologies into contexts of use so that we might learn how people appropriate the technologies of shape change going beyond the surprising moments of the first encounters.

In terms of cognitive robotics and interaction design, a recent paper draws attention to the rise of ever more complex objects, and environments in which there are not clearly demarcated regions on a surface or clearly defined buttons to push. Going beyond things and beings, they proposed 'fields' of interaction in which the human inhabits a space with interactive forces and possibilities [18]. How should these complex objects, a mix of entities and environments share information, build a memory, learn from the past? How can robots develop more meaningful expressions with shape change to support the human in such an environment?

5.7   ACKNOWLEDGMENTS

We thank all of the students from the Aarhus University IT Product Development Master of Science program for their generous contributions and for sharing details of their work and design processes.

5.8 Bibliography

[1] Tei 2014. Call for Arts Track Submissions - TEI 2014, Munich, Germany, 2014.

[2] Edith K. Ackermann. Playthings That Do Things: A Young Kid's "Incredibles"! In *Proceedings of the 2005 Conference on Interaction Design and Children*, IDC '05, pages 1–8, New York, NY, USA, 2005. ACM.

[3] Tobias Alrøe, Jonas Grann, Erik Grönvall, Marianne Graves Petersen, and Jesper L. Rasmussen. Aerial tunes: Exploring interaction qualities of mid-air displays. In *Proceedings of the 7th Nordic Conference on Human-Computer Interaction: Making Sense Through Design*, NordiCHI '12, pages 514–523, New York, NY, USA, 2012. ACM.

[4] Tilde Bekker, Janienke Sturm, Rik Wesselink, Bas Groenendaal, and Berry Eggen. Interactive Play Objects and the Effects of Open-ended Play on Social Interaction and Fun. In *Proceedings of the 2008 International Conference on Advances in Computer Entertainment Technology*, ACE '08, pages 389–392, New York, NY, USA, 2008. ACM.

[5] Bill Buxton. *Sketching User Experiences: Getting the Design Right and the Right Design (Interactive Technologies)*. Morgan Kaufmann, 1 edition, San Francisco, April 2007.

[6] Marcelo Coelho and Pattie Maes. Shutters: A Permeable Surface for Environmental Control and Communication. In *Proceedings of the 3rd International Conference on Tangible and Embedded Interaction*, TEI '09, pages 13–18, New York, NY, USA, 2009. ACM.

[7] Marcelo Coelho and Jamie Zigelbaum. Shape-changing Interfaces. *Personal Ubiquitous Comput.*, 15(2):161–173, February 2011.

[8] Hugo Fastl. Psycho-acoustics and sound quality. In *Communication acoustics*, pages 139–162. Springer, Berlin, 2005.

[9] Kieran Ferris and Liam Bannon. "...A Load of Ould Boxology!". In *Proceedings of the 4th Conference on Designing Interactive Systems: Processes, Practices, Methods, and Techniques*, DIS '02, pages 41–49, New York, NY, USA, 2002. ACM.

[10] Sean Follmer, Daniel Leithinger, Alex Olwal, Nadia Cheng, and Hiroshi Ishii. Jamming User Interfaces: Programmable Particle Stiffness and Sensing for Malleable and Shape-changing Devices. In *Proceedings of the 25th Annual ACM Symposium on User Interface Software and Technology*, UIST '12, pages 519–528, New York, NY, USA, 2012. ACM.

[11] Christopher Frayling. *Research in art and design*. Royal College of Art London, 1993.

[12] Johanna Fulda, Pieter Tierens, Teemu Mäntyharju, and Thomas Wimmer. Raising User Impulse Awareness: The Sensitive RolyPoly. In *Proceedings of the 8th International Conference on Tangible, Embedded and Embodied Interaction*, TEI '14, pages 347–348, New York, NY, USA, 2013. ACM.

[13] Erik Grönvall, Sofie Kinch, Marianne Graves Petersen, and Majken K. Rasmussen. Causing commotion with a shape-changing bench: Experiencing shape-changing interfaces in use. In *Proceedings of the SIGCHI Conference on Human Factors in Computing Systems*, CHI '14, pages 2559–2568, New York, NY, USA, 2014. ACM.

[14] Chris Harrison and Scott E. Hudson. Providing Dynamically Changeable Physical Buttons on a Visual Display. In *Proceedings of the SIGCHI Conference on Human Factors in Computing Systems*, CHI '09, pages 299–308, New York, NY, USA, 2009. ACM.

[15] Fabian Hemmert, Susann Hamann, Matthias Löwe, Josefine Zeipelt, and Gesche Joost. Shape-changing Mobiles: Tapering in Two-dimensional Deformational Displays in Mobile Phones. In *CHI '10 Extended Abstracts on Human Factors in Computing Systems*, CHI EA '10, pages 3075–3080, New York, NY, USA, 2010. ACM.

[16] Fabian Hemmert, Matthias Löwe, Anne Wohlauf, and Gesche Joost. Animate Mobiles: Proxemically Reactive Posture Actuation As a Means of Relational Interaction with Mobile Phones. In *Proceedings of the 7th International Conference on Tangible, Embedded and Embodied Interaction*, TEI '13, pages 267–270, New York, NY, USA, 2013. ACM.

[17] Hiroshi Ishii, Dávid Lakatos, Leonardo Bonanni, and Jean-Baptiste Labrune. Radical atoms: Beyond tangible bits, toward transformable materials. *interactions*, 19(1):38–51, January 2012.

[18] Lars-Erik Janlert and Erik Stolterman. Faceless interaction - a conceptual examination of the notion of interface: past, present and future. *Human Computer Interaction*, 0(ja):null, 2014.

[19] Yuichiro Katsumoto, Satoru Tokuhisa, and Masa Inakage. Ninja Track: Design of Electronic Toy Variable in Shape and Flexibility. In *Proceedings of the 7th International Conference on Tangible, Embedded and Embodied Interaction*, TEI '13, pages 17–24, New York, NY, USA, 2013. ACM.

[20] Hyunjung Kim. Designing Interactive Kinetic Surfaces for Everyday Objects and Environments. In *Proceedings of the Fourth International Conference on Tangible, Embedded, and Embodied Interaction*, TEI '10, pages 301–302, New York, NY, USA, 2010. ACM.

[21] Naoya Koizumi, Kentaro Yasu, Angela Liu, Maki Sugimoto, and Masahiko Inami. Animated Paper: A Toolkit for Building Moving Toys. *Comput. Entertain.*, 8(2), December 2010.

[22] Jakob Leitner, Michael Haller, Kyungdahm Yun, Woontack Woo, Maki Sugimoto, and Masahiko Inami. IncreTable, a Mixed Reality Tabletop Game Experience. In *Proceedings of the 2008 International Conference on Advances in Computer Entertainment Technology*, ACE '08, pages 9–16, New York, NY, USA, 2008. ACM.

[23] Alex Mitchell and Kevin McGee. Limits of Rereadability in Procedural Interactive Stories. In *Proceedings of the SIGCHI Conference on Human Factors in Computing Systems*, CHI '11, pages 1939–1948, New York, NY, USA, 2011. ACM.

[24] Jules Moloney. Between art and architecture: The interactive skin. In *Information Visualization, 2006. IV 2006. Tenth International Conference on*, pages 681–686. IEEE, London, 2006.

[25] Ditte Hvas Mortensen, Sam Hepworth, Kirstine Berg, and Marianne Graves Petersen. "It's in love with you": Communicating status and preference with simple product movements. In *CHI '12 Extended Abstracts on Human Factors in Computing Systems*, CHI EA '12, pages 61–70, New York, NY, USA, 2012. ACM.

[26] Bilge Mutlu, Fumitaka Yamaoka, Takayuki Kanda, Hiroshi Ishiguro, and Norihiro Hagita. Nonverbal leakage in robots: Communication of intentions through seemingly unintentional behavior. In *Proceedings of the 4th ACM/IEEE International Conference on Human Robot Interaction*, HRI '09, pages 69–76, New York, NY, USA, 2009. ACM.

[27] Kristina Niedderer. Exploring elastic movement as a medium for complex emotional expression in silver design. *International Journal of Design*, 6(3):57–69, 2012.

[28] Mie Nørgaard, Tim Merritt, Majken K. Rasmussen, and Marianne G. Petersen. Exploring the Design Space of Shape-Changing Objects: Imagined Physics. In *Proceedings of the 6th International Conference on Designing Pleasurable Products and Interfaces*, DPPI '13, pages 251–260, New York, NY, USA, 2013. ACM.

[29] Kas Oosterhuis and Nimish Biloria. Interactions with Proactive Architectural Spaces: The Muscle Projects. *Commun. ACM*, 51(6):70–78, June 2008.

[30] Leonhard Oschuetz, Daniel Wessolek, and Wolfgang Sattler. Constructing with Movement: Kinematics. In *Proceedings of the Fourth International Conference on Tangible, Embedded, and Embodied Interaction*, TEI '10, pages 257–260, New York, NY, USA, 2010. ACM.

[31] Amanda Parkes and Hiroshi Ishii. Bosu: A Physical Programmable Design Tool for Transformability with Soft Mechanics. In *Proceedings of*

the 8th ACM Conference on Designing Interactive Systems, DIS '10, pages 189–198, New York, NY, USA, 2010. ACM.

[32] Amanda Parkes, Ivan Poupyrev, and Hiroshi Ishii. Designing kinetic interactions for organic user interfaces. *Commun. ACM*, 51(6):58–65, June 2008.

[33] Amanda Jane Parkes. *Phrases of the kinetic: dynamic physicality as a dimension of the design process*. PhD thesis, Massachusetts Institute of Technology, 2009.

[34] Christine Patsouras, Hugo Fastl, Ulrich Widmann, and Georg Hölzl. Psychoacoustic evaluation of tonal components in view of sound quality design for high-speed train interior noise. *Acoustical Science and Technology*, 23(2):113–116, 2002.

[35] Esben W. Pedersen, Sriram Subramanian, and Kasper Hornbæk. Is my phone alive?: A large-scale study of shape change in handheld devices using videos. In *Proceedings of the 32Nd Annual ACM Conference on Human Factors in Computing Systems*, CHI '14, pages 2579–2588, New York, NY, USA, 2014. ACM.

[36] Søren Pedersen, Michael Ha, Christian Ø. Laursen, and Anders Høedholt. Kinetic Wave: Raising Awareness of the Electromagnetic Spectrum. In *Proceedings of the 8th International Conference on Tangible, Embedded and Embodied Interaction*, TEI '14, pages 337–338, New York, NY, USA, 2013. ACM.

[37] Ivan Poupyrev, Tatsushi Nashida, and Makoto Okabe. Actuation and tangible user interfaces: The Vaucanson duck, robots, and shape displays. In *Proceedings of the 1st International Conference on Tangible and Embedded Interaction*, TEI '07, pages 205–212, New York, NY, USA, 2007. ACM.

[38] Kathrin Probst, Thomas Seifried, Michael Haller, Kentaro Yasu, Maki Sugimoto, and Masahiko Inami. Move-it: Interactive Sticky Notes Actuated by Shape Memory Alloys. In *CHI '11 Extended Abstracts on Human Factors in Computing Systems*, CHI EA '11, pages 1393–1398, New York, NY, USA, 2011. ACM.

[39] Hayes S. Raffle, Amanda J. Parkes, and Hiroshi Ishii. Topobo: A Constructive Assembly System with Kinetic Memory. In *Proceedings of the SIGCHI Conference on Human Factors in Computing Systems*, CHI '04, pages 647–654, New York, NY, USA, 2004. ACM.

[40] Majken K. Rasmussen, Esben W. Pedersen, Marianne G. Petersen, and Kasper Hornbaek. Shape-changing Interfaces: A Review of the Design

Space and Open Research Questions. In *Proceedings of the SIGCHI Conference on Human Factors in Computing Systems*, CHI '12, pages 735–744, New York, NY, USA, 2012. ACM.

[41] Susanne Seitinger, Elisabeth Sylvan, Oren Zuckerman, Marko Popovic, and Orit Zuckerman. A New Playground Experience: Going Digital? In *CHI '06 Extended Abstracts on Human Factors in Computing Systems*, CHI EA '06, pages 303–308, New York, NY, USA, 2006. ACM.

[42] Janienke Sturm, Tilde Bekker, Bas Groenendaal, Rik Wesselink, and Berry Eggen. Key Issues for the Successful Design of an Intelligent, Interactive Playground. In *Proceedings of the 7th International Conference on Interaction Design and Children*, IDC '08, pages 258–265, New York, NY, USA, 2008. ACM.

[43] Yuta Sugiura, Calista Lee, Masayasu Ogata, Anusha Withana, Yasutoshi Makino, Daisuke Sakamoto, Masahiko Inami, and Takeo Igarashi. PINOKY: A Ring That Animates Your Plush Toys. In *Proceedings of the SIGCHI Conference on Human Factors in Computing Systems*, CHI '12, pages 725–734, New York, NY, USA, 2012. ACM.

[44] Gordon Tiemstra, Renée Van Den Berg, Tilde Bekker, and MJ de Graaf. Guidelines to design interactive open-ended play installations for children placed in a free play environment. In *Proceedings of the international conference of the digital research association (DIGRA). Utrecht, The Netherlands*, pages 1–17, 2011.

[45] Anna Vallgaarda. The dress room: Responsive spaces and embodied interaction. In *Proceedings of the 8th Nordic Conference on Human-Computer Interaction: Fun, Fast, Foundational*, NordiCHI '14, pages 618–627, New York, NY, USA, 2014. ACM.

[46] Anna Vallgaarda and Johan Redström. Computational composites. In *Proceedings of the SIGCHI Conference on Human Factors in Computing Systems*, CHI '07, pages 513–522, New York, NY, USA, 2007. ACM.

[47] John P Whitney, Matthew F Glisson, Eric L Brockmeyer, and Jessica K Hodgins. A low-friction passive fluid transmission and fluid-tendon soft actuator, pages 2801–2808, Chicago, 2014.

[48] Jamie Zigelbaum, Angela Chang, James Gouldstone, Joshua J. Monzen, and Hiroshi Ishii. SpeakCup: Simplicity, BABL, and Shape Change. In *Proceedings of the 2nd International Conference on Tangible and Embedded Interaction*, TEI '08, pages 145–146, New York, NY, USA, 2008. ACM.

[49] John Zimmerman, Jodi Forlizzi, and Shelley Evenson. Research through design as a method for interaction design research in hci. In *Proceedings of the SIGCHI Conference on Human Factors in Computing Systems*, CHI '07, pages 493–502, New York, NY, USA, 2007. ACM.

V

Cultural & Social Aspects of Cognitive Robotics

CHAPTER 6

Effects of Cultural Context and Social Role on Human–Robot Interaction

Pei-Luen Patrick Rau

Department of Industrial Engineering,
Tsinghua University, China.

Na Chen

Department of Industrial Engineering,
Tsinghua University, China.

CONTENTS

6.1 INTRODUCTION

R ECENTLY, the technologies of robotics have been extended from majorly industrial fields to health care, social services, aerospace and so on. The role of robots, in particular, is moving from a traditional one of supplement of labour shortage and solution of dangerous work to becoming an important part of users' daily lives - social robots. Social robots provide users life service such as house cleaning and babysitting, as well as decision-making support. Previous findings indicated that participants performed social behaviours while in contact with humanoid robots [4]. Tmsuk developed shopping robots, in place of humans, making purchase decisions based on information collected in markets [10]. iRobot - one of the most famous robotics companies - had sold more than 10 million home robots by 2013.

Recent studies of human-robot interaction (HRI) have revealed the influences of design factors of robotics on human decision-making. These factors include physical characteristics of robots (i.e. appearance, size, humanoid or machine-like, etc.), methods of information interaction (i.e. gesture, vision, language, etc.), training strategies (i.e. showing and teaching, programming, etc.), structures of robot user teams (i.e. the roles of users - cooperators, competitors, or operators), task design and so on. There should be more work on the influences of the interaction environments, interaction tasks, robots' attributes and the background and experience of users [23]. This chapter reviews HRI studies in respect to the following aspects: cultural context, robots' social role and their effects on human decision-making. The direction of future work in the field of cross-cultural HRI and decision-making is discussed.

6.2 EFFECTS OF CULTURAL CONTEXT

6.2.1 Factors Influencing Human–Robot Interaction

There are four types of factors which should be considered in the research of HRI, i.e. factors related to human beings, factors related to robotics, factors related to tasks and factors related to contexts.

The factors related to robotics include forms and appearance, behaviours, learning ability, characteristics, affection, communication methods, appearance during the experiment and purposes [9]. When the robots are human-shaped, the factors related to robotics also include gender, nationality and facial characteristics. The appearance of robots also influences human perception [7, 11, 14, 27], including how humans assume robots' abilities, how humans behave with robots and their judgement on robots. Mashiro Mori proposed a famous model - uncanny valley [38]. Mori stated that when the appearance of a robot is made more humanoid, participants will respond towards the robot more positively. When the similarity increases and reaches a point, the response quickly becomes strong revulsion. However, as the robot's appearance continues to become more humanoid, participants' responses become positive once again and finally approach the level of empathy between human and human.

The factor of whether or not robots appear during the experiment has a significant influence on human perception [16], including how humans judge robots' altruism and their trust of robots. Previous researchers argued that when robots appeared on the scene, participants showed higher trust of robots and thought the robots were more altruistic. Compared with the animated characters created by computers, robots have more social attributes, are more real and can share spaces with humans. Some researchers argued that robots' behaviours influence the interaction between human beings and robots. The eye contact between human beings and robots [15], eye tracking of robots [38] and the cooperative behaviours between human beings and robots would improve human evaluations of robots. The mental model of robots is said to be influenced by the sources and languages of robots [19].

Based on the research of Lee, Park and Song, robots' ability of self-development is said to influence participants' cognition, especially the distinguishing of robots' social attributes [18]. Although in the experiment, the learning process was simplified, it indeed proposed a new view about how to design robots' abilities. Future robots are expected to have the ability to develop the self as time goes on in order to adapt to the environment better.

The factors related to humans include participants' genders, ages, characteristics, background, the experience of interaction with robots and knowledge of robots [24-26, 31, 37]. Cultural differences will also influence human cognition and evaluation of robots as well as behaviours towards robots. Based on the results of an experiment conducted with participants from six countries (i.e. Dutch, Chinese, German, Mexican, American and Japanese), the American participants were least negative towards robots; the Mexican participants were most negative; and the Japanese participants did not have particularly positive attitudes towards robots [35]. Additionally, the experience of interaction with robots was said to have a positive influence on participants' attitudes.

6.2.2 Effects of Cultural Context on Decision Making

In 2007, Rau and his team carried out a study to investigate the influences of participants' cultural background (human-related factor), the appearance of robots (robot-related factor) and robots' tasks (task-related factor) on the interaction between human beings and robots [20-21]. The independent variables of participants' cultural background and the appearance of robots included three levels (Chinese, Korean, and German; zoomorphic, humanoid and machine-like) and were between-subject variables. The independent variables of robots' tasks included four levels (teaching, guiding, entertaining and security) and were within-subject variables. The influences include two aspects - preference (likability) and interaction performance (active response, engagement, trust and satisfaction).

During the experiment, participants were required to talk with the robots in their native languages and finish some tasks following the robots' instructions. The experiment tasks were designed based on the functions existing robots had or would have. The tasks of teaching included English teaching and health-care teaching. In the tasks of guiding, robots guided the participants to visit the former residence of Copernicus. The entertaining tasks included playing music and telling jokes. The security tasks included instructing on secure functions and performing security measures with respect to buildings.

Participants did not need to operate the robots by themselves. The interaction was initiated by the robots. The interaction between participants and robots included oral interaction and behavioural interaction. The robots just performed scattered, short-term and basic behaviours in order to avoid the influence of their body language on participants. For the humanoid robots, the behaviours included shaking heads, raising arms and moving legs. For the machine-like robots, the behaviours included moving forwards and backwards through the rotation of wheels. For the zoomorphic robots, the behaviours just included nodding. The movements of all three types of robots were designed through the platform of Lego Mindstorm NXT. In total, 108 participants were recruited in this experiment, including 36 participants from Chinese, Korean and German cultures respectively. According to the results, Chinese and Korean participants showed higher likability, engagement and satisfaction than German participants. When the interaction mainly relied on oral communication rather than body language and environmental support, the engagement of Chinese and German participants was lower than Korean participants in all contexts.

In the same year of 2007, Rau's team conducted another experiment to investigate the impacts of robots' recommendations on human decision-making behaviours [22, 28]. The experiment included a series of price-guessing tasks. Participants were required to guess which one was more expensive from two similar items. After participants made the primary decision, robots provided their suggestions. Participants had a choice to change or not to change their first decision. This study considered two aspects: two robot-related variables (i.e. robots' communication styles and language) and one human-related variable (participants' cultural background). 16 Chinese and 16 German participants were recruited to take part in the experiment. In different contexts, the robots communicated with the participants through different methods - implicit or explicit recommendations. The results indicated that Chinese participants considered robots more likable, more trustable and more credible than German participants. When making decisions, Chinese participants were more likely to be influenced by robots than German participants. Both Chinese and German participants preferred to communicate with robots using implicit recommendations rather than explicit recommendations.

6.3 EFFECTS OF ROBOTS' SOCIAL ROLE: AUTONOMY LEVEL AND GROUP ORIENTATION

6.3.1 Social Robots

The rapid development of robotics technologies leads to the continuous extension of applications of robotics. With the popularity of service robots, more and more robots appear in public service and home support contexts. There are generally two types of service robots: instrumental robots and social robots [33]. Instrumental robots have advantages over industrial robots and provide humans with a convenient life. For example, the Roomba vacuum-cleaning robot produced by iRobot can help users clean rooms [13]; the hospital robots produced by Panasonic can help nurses deliver food, medicine, documents, instruments and even patients [1]. According to the demands of functions, instrumental robots are usually designed in a machine-like manner. The user interfaces of instrumental robots are designed with the most effective and direct methods to receive and transfer information. There will be no interactive feedback from this type of robot.

Social robots are a new application of robots. To interact at close range with users is one of the major purposes of this type of robot. Compared with industrial robots and instrumental robots such as service robots, social robots are more autonomous and communicative and are more often used in entertainment, education, health care, research and development, and public places [2].

While communicating with others, users will display social behaviours naturally, such as talking in natural languages, transferring information by body language, maintaining interpersonal distance and so on. Previous research indicated that while communicating with artificial objects with enough intelligence and complexity or those objects which can represent the real world, users will also display social behaviours. For example, observers will think the characters shown on TV or the virtual people generated by computers have social attributes, so the observers perform social behaviours with them [30]. When social robots have enough social attributes, they will encourage users to perform social behaviours with them in order to realise social interaction.

6.3.2 Autonomy Level of Robots

The autonomy of robots is a major design factor, which will influence users' engagement, mental workload, psychological cognition of the interactive objects, the performance level and so on. Among the various standards which evaluate robots' autonomy, autonomy level is a commonly used standard in used-centred design. Autonomy level describes to which degree robots can act according to established procedures or patterns [23]. Sheridan and Verplank [32] developed a scale to explicitly compare different mixes of human and robot decision-making. This scale includes ten levels of autonomy going from the lowest level wherein the human operator does everything to a level wherein the robot does everything. The ten levels are described in the following manner (we replaced "computer" in the original book with "robot"):

1) human does everything up to the point of turning it over to the robot to implement;

2) robot helps by determining the options;

3) robot helps determine options and makes suggestions, which the human need not follow;

4) robot selects action and human may or may not do it;

5) robot selects action and implements it if human approves;

6) robot selects action but informs human with plenty of time to stop it;

7) robot does whole job and necessarily tells human what it did;

8) robot does whole job and tells human what it did only if human explicitly asks;

9) robot does whole job and tells human what it did if it decides he/she should be told;

10) robot does whole job if it decides it should be done, and if so tells human if it decides he/she should be told.

The research of Siegel [33] from MIT Media Lab provided some support for the effects of robots' autonomy level on their persuasion ability. In their experiment, participants were made to trust that the robots they were cooperating with 1) had a full autonomy level or 2) were operated by an experimenter remotely. During the experiment, robots tried to persuade participants to donate more money to them. According to the results, while cooperating with the robots, which were believed to have no autonomy (controlled by an experimenter remotely), participants would donate more money. The female participants thought the robots with a full autonomy level were more reliable, and at the same they showed higher engagement, while the male participants rated the fully autonomous robots lower.

Because of the current application of social robots, there are two major control methods of robots in experiments—controlled by operators or by pre-programmed procedures. For both control methods, it is not common to expose participants fully to the control mechanism. In real-life applications, it is not easy to control how participants perceive the autonomy level of the robots. The perception of robots' autonomy level varies largely with participants' background, experience of robots and knowledge levels.

6.3.3 Group Orientation

Group orientation includes in-group orientation and out-group orientation. Triandis and his colleagues [36] defined a commonly referred conception of in-group relations - those relations people are required to maintain. More specifically, people care about the welfare of in-group members, people try to cooperate with in-group members and do not ask for equal returns. Even if somebody else in the group makes them feel uncomfortable, people won't leave the group. On the other hand, out-group relations are defined as those relations people are not required to maintain - people do not care about the welfare of out-group members and people ask for equivalent returns while cooperating with out-group members [12].

Research has indicated that the same group orientation could improve the cooperation of in-group members and this cooperation intention originated from in-group favouritism and positive evaluation of in-group members [3, 5, 17]. The positive cooperation of in-group members would emerge through grouping by task-unrelated factors. The research of Terry and Hogg [34] further indicated that the participants who had a stronger sense of belonging to groups were more easily influenced by other in-group members if they were grouped by random factors.

To encourage humans and robots to form in-group relations, Evers, Maldonado, Brodecki and Hinds [8] conducted a survey to find the most important factor which aided in forming in-group/out-group cognition - long history of interaction with interactive objects and experience of successful cooperation. In the following experiment, the researchers developed two treatment levels of this factor. In the experiment groups, participants and robots built up in-group relations, while in the control groups, participants and robots did not build up in-group relations. In the post-experiment questionnaire, there was no significant difference in cognition of relations toward in-group robots between participants in the experiment groups and those in the control groups.

Another experiment of HRI also did not find any differences in the relations with in-group robots [39]. In this experiment, signs of different colours were used to distinguish in-group/out-group relations. There was no significant difference in different coloured sign settings while participants interacted with robots. However, while participants interacted with humans, significant differences were found among different settings.

Another experiment indicated that when participants considered a robot as their in-group member, they would rate the robot as reliable and trustable, and they more easily accepted the robot's recommendations [40]. The result of this experiment showed that in-group favouritism also existed in human-robot interactive groups. However, it should be noted that the methods of building human-robot interactive groups might be different from the methods of group building among humans.

6.3.4 Effects of Robots' Social Role on Decision Making

Based on the price-guess experiment in 2007 mentioned above, Rau's team improved the experiment tasks to investigate the effects of social roles of social robots (autonomy level and group orientation) in HRI [23, 29]. Both quantitative psychological questionnaire and qualitative observation were used to record participants' feedback. Additionally, participants' gender,

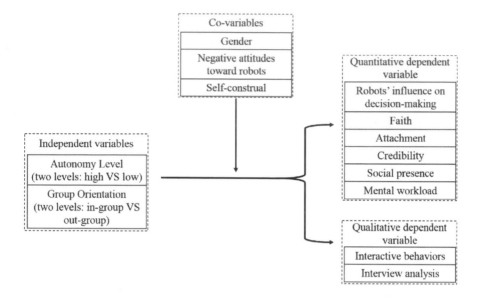

FIGURE 6.1 Study framework of robots' social role in Rau's 2010 HRI experiment.

self-construal, negative attitude towards robots as well as accuracy initial decisions were chosen as the covariates. Robots' influence on participants' decision-making, faith, attachment, credibility, social presence, mental workload and interactive behaviours were chosen as the independent variables.

While choosing experimental tasks, three aspects should be taken into consideration: 1) the tasks should be cooperative tasks which require participants and robots to make decisions together; 2) the tasks should represent the real world; 3) the experiment should be realised easily within the current experimental platform. The cooperation tasks ensure that the participants do not have absolute advantages in decision-making and action capability, beyond the abilities of robots, to conduct the tasks. Both participants and robots have independent decision-making and action capability. In the real application, social robots which act as decision-making supporters can have roles as experts, advisers, service personnel, partners and so on to provide users information and suggestions. This type of robot should not only have the same ability as general social robots to encourage users' social behaviours, but also have the ability to build trust with users. In this experiment, robots acted as participants' partners and provided decision-making support. When considering all issues, the tasks of sea survival from the US Army Survival Manual [6] were chosen as the task prototype. The study framework is outlined in Figure 6.1.

The task scenario could be described as follows. The participant and the robot are sailing a boat on the Atlantic Ocean. There happens to be a serious fire so the participant and the robot must abandon the boat. To return to land safely, the participant and the robot should make a series of decisions: 1) while abandoning the boat, select the six most important items for survival to take to the lifeboat from a total of twelve items and take the six items; 2) while operating the lifeboat, decide about how to set up the sails and lay the anchor, and choose the methods to avoid the sharks and the locate land.

In total, 48 participants took part in the experiment. Those who had ever participated in an experiment using this robot were excluded from recruitment. Based on the results, participants were more likely to accept recommendations from the robot with a higher level of autonomy, showed higher attachment to the robot and glared for a longer amount of time at the robot. Participants had a lower mental workload while cooperating with the in-group robot. Compared with cooperation with the out-group robot, the robots' autonomy level showed more influence on participants' acceptance of recommendations while they cooperated with the in-group robot. The female participants thought the robots were more reliable than the male participants did and shared more benefits with robots. The participants who showed higher negative attitudes towards robots, measured before experiment, had a higher mental workload.

6.4   FUTURE WORK

As the popularity of social robots in public places and families increases, users' cognition and reflection towards the robots will show a higher level of predictability. The influences of robot design factors on users' behaviours and cognition will become more clear. Currently, there is still a lack of research targeted at how the real interaction experience of users influences their real interactive behaviours with robots and decision-making behaviours. The theoretical framework of the interaction experiment and interactive behaviour has not been built. Research investigating the influences of users' interaction experience with robots on user cognition and behaviours is valuable. This research can improve robot designers' abilities to predict user experience and behaviours, in order to develop robots which can maintain positive user experience.

There are two directions in which to further investigate the influences of users' interaction experience with robots. The first is to screen the users who have different levels of interaction experience. The second is to require participants to interact with robots several times during a long period, in order to measure the influence of the increase of interaction experience on the participants' cognition and behaviours.

Additionally, considering the demand that social robots interact with multiple users at the same time, the research of group decisions supported by robots is another future research direction.

6.5 BIBLIOGRAPHY

1. E. Ackerman. Panasonic revives hospital delivery robot. Retrieved September 1, 2014, from http://spectrum.ieee.org/automaton/robotics/medical-robots/panasonic-hospital-delivery-robot.

2. C. L. Breazeal. Designing Sociable Robots. Cambridge, MA: MIT Press, 2004.

3. M. B. Brewer. In-group bias in the minimal intergroup situation: A cognitive-motivational analysis. Psychological Bulletin, 86(2):307-324, 1979.

4. R. Byron and C. Nass. How People Treat Computers, Television, and New Media Like Real People and Places. Stanford, CA: CSLI Publications and Cambridge University Press, 1996.

5. Y. Chen, J. Brockner, and X. Chen. Individual-collective primacy and ingroup favoritism: Enhancement and protection effects. Journal of Experimental Social Psychology, 38(5):482-491, 2002.

6. Department of Defense, the US Army. Chapter 16: Sea survival. In US Army Survival Manual, pages 21-76. Skyhorse Publishing Inc., New York, NY, 1992.

7. C. F. DiSalvo, F. Gemperle, J. Forlizzi, et al. All robots are not created equal: The design and perception of humanoid robot heads. In Proceedings of the 4th Conference on Designing Interactive Systems: Processes, Practices, Methods, and Techniques, pages 321-326. ACM, 2002.

8. V. Evers, H. Maldonado, T. Brodecki, and P. Hinds. Relational vs. group self-construal: Untangling the role of national culture in HRI. In Proceedings of the 3rd ACM/IEEE International Conference on Human-Robot Interaction (HRI), pages 255-262. IEEE, 2008.

9. T. Fong, I. Nourbakhsh, and K. Dautenhahn. A survey of socially interactive robots. Robotics and Autonomous Systems, 42(3):143-166, 2003.

10. G. Galitzine. If you see a robot in the mall. Retrieved August 15, 2014, from http://blog.tmcnet.com/blog/greg-galitzine/voip/robotics/if-you-see-a-robot-in-the-mall.html.

11. J. Goetz, S. Kiesler, and A. Powers. Matching robot appearance and behavior to tasks to improve human-robot cooperation. In Proceedings of the 12th IEEE International Workshop on Robot and Human Interactive Communication, pages 55-60. IEEE, 2003.

12. W. B. Gudykunst and M. H. Bond. Intergroup relations across cultures. Handbook of Cross-Cultural Psychology, 3:119-161, 1997.

13. iRobot Corporation. Roomba - Vacuum cleaning robot. Retrieved August 15, 2014, from http://store.irobot.com/family/index.jsp?categoryId=2501652.

14. H. Ishiguro. Scientific issues concerning androids. The International Journal of Robotics Research, 26(1):105-117, 2007.

15. T. Kanda, H. Ishiguro, and T. Ishida. Psychological analysis on human-robot interaction. In Proceedings of the IEEE International Conference on Robotics and Automation, pages 4166-4173. IEEE, 2001.

16. C. D. Kidd. Sociable Robots: The Role of Presence and Task in Human-Robot Interaction. Dissertation. Massachusetts Institute of Technology, 2003.

17. R. M. Kramer and L. Goldman. Helping the group or helping yourself? Social motives and group identity in resource dilemmas. In Social dilemmas: Perspectives on Individuals and Groups: Greenwood Publishing Group, Santa Barbara, CA, 49-67, 1995.

18. K. Lee, N. Park, and H. Song. Can a robot be perceived as a developing creature? Human Communication Research, 31(4):538-563, 2005.

19. S. Lee, I. Lau, S. Kiesler, and C. Chiuet. Human mental models of humanoid robots. In Proceedings of the 2005 IEEE International Conference on Robotics and Automation, pages 2767-2772. IEEE, 2005.

20. D. Li. Affect of Appearance, Task and Culture in Human Robot Interaction. Dissertation. Tsinghua University, 2008.

21. D. Li, P. L., Rau, & Y. Li. A cross-cultural study: effect of robot appearance and task. International Journal of Social Robotics, 2(2): 175-186, 2010.

22. Y. Li. The Effects of Robots' Recommendation on Human's Decision Making. Dissertation. Tsinghua University, Beijing, China, 2007.

23. Y. Li. The Effect of a Social Robot's Autonomy and Group Orientation on Human Decision-Making. Dissertation. Tsinghua University, Beijing, China, 2010.

24. B. Mutlu, S. Osman, J. Forlizzi, et al. Task structure and user attributes as elements of human-robot interaction design. In Proceedings of the 15th IEEE International Symposium on Robot and Human Interactive Communication, pages 74-79. IEEE, 2006.

25. T. Nomura, T. Kanda, and T. Suzuki. Experimental investigation into influence of negative attitudes toward robots on human-robot interaction. AI & Society, 20(2):138-150, 2006.

26. A. Powers and S. Kiesler. The advisor robot: Tracing people's mental model from a robot's physical attributes. In Proceedings of the 1st ACM SIGCHI/SIGART Conference on Human-Robot Interaction, pages 218-225. ACM, 2006.

27. T. Nomura, T. Tasaki, T. Kanda, et al. Questionnaire-based research on opinions of visitors for communication robots at an exhibition in Japan. In Proceedings of Human-Computer Interaction - Interact 2005, pages 685-698. Springer, Berlin, Heidelberg, 2005.

28. P. L. Rau, Y. Li, & D. Effects of communication style and culture on ability to accept recommendations from robots. Computers in Human Behavior, 25(2): 587-595, 2009.

29. P. L. Rau, Y. Li, & J. Liu. Effects of a social robot's autonomy and group orientation on human decision-making. Advances in Human-Computer Interaction, 11:1-13, 2013.

30. B. Reeves and C. Nass. The media equation: How people treat computers, television and new media like real people and places. New York, NY: Cambridge University Press, 1997.

31. M. Scopelliti, M. V. Giuliani, and F. Fornara. Robots in a domestic setting: A psychological approach. Universal Access in the Information Society, 4(2):146-155, 2005.

32. T. B. Sheridan and W. L. Verplank. Human and Computer Control of Undersea Teleoperators. Massachusetts Institute of Technology Man-Machine Systems Lab, Cambridge, MA, USA, 1978.

33. M. S. Siegel. Persuasive Robotics: How Robots Change Our Minds. Dissertation. Massachusetts Institute of Technology, 2008.

34. D. J. Terry and M. A. Hogg. Group norms and the attitude-behavior relationship: A role for group identification. Personality and Social Psychology Bulletin, 22(8):776-793, 1996.

35. T. Tojo, Y. Matsusaka Y, T. Ishii, and T. Kobayashi. A conversational robot utilizing facial and body expressions. In IEEE International Conference on Systems, Man, and Cybernetics, pages 858-863. IEEE, 2000.

36. H. C. Triandis, R. Bontempo, M. J. Villareal, et al. Individualism and collectivism: Cross-cultural perspectives on self-ingroup relationships. Journal of Personality and Social Psychology, 54(2):323-338, 1988.

37. M. L. Walters, K. Dautenhahn, K. L. Koay, et al. The influence of subjects' personality traits on predicting comfortable human-robot approach distances. In Proceedings of the Cog Sci 2005 Workshop: Toward Social Mechanisms of Android Science, pages 29-37. Cog Sci, 2005.

38. E. Wang, C. Lignos, A. Vatsal, and B. Scassellati. Effects of head movement on perceptions of humanoid robot behavior. In Proceedings of the 1st ACM SIGCHI/SIGART Conference on Human-Robot Interaction, pages 180-185. ACM, 2006.

39. L. Wang, P. P. Rau, V. Evers, et al. When in Rome: The role of culture & context in adherence to robot recommendations. In Proceedings of the 5th ACM/IEEE International Conference on Human-Robot Interaction, pages 359-366. IEEE, 2010.

40. L. Wang, P. P. Rau, V. Evers V, et al. Responsiveness to robots: Effects of ingroup orientation & communication style on HRI in China. In Proceedings of the 4th ACM/IEEE International Conference on Human-Robot Interaction (HRI), pages 247-248. IEEE, 2009.

CHAPTER 7

A Cognitive Model for Human Willingness to Collaborate with Robots: The Emergence of Cultural Robotics

Belinda J. Dunstan

Creative Robotics Lab,
University of New South Wales Art and Design, Australia

Jeffrey Tzu Kwan Valino Koh

Creative Robotics Lab,
University of New South Wales Art and Design, Australia

CONTENTS

THIS chapter identifies the emergence of a desire amongst contemporary social roboticists for increased collaboration, deeper interaction and more meaningful and subtle relationships in human-robot interaction. Therefore we contribute a cognitive model that has been developed to identify the determinants of human willingness to collaborate with a robot counterpart, and specifically what factors influence our assessment of our own capacity and that of a robot prior to and during collaboration. We argue that the assessment of a robot will be fast, automatic and based primarily upon an ocularcentric judgement of aesthetics, and the culture of both the human participant and that which has shaped and is ultimately projected by, the robot. With this in mind we introduce the emergence of "Cultural Robotics" as consideration for the future development of collaborative robotic agents.

7.1   INTRODUCTION

In the last decade, the field of social robotics has seen rapid growth and the development of many variants of 'sociable machines' [16]. With functions ranging from museum tour guides [60] to bartenders [31], social robots have been developed to participate in an extraordinary range of roles, including domestic helpers [58], musical collaborators [63], and friends [62]. These 'relational artefacts' no longer simply do things for us, they do things with us [62].

The increasing integration of robots into human life requires adaptations to be made, not simply technical advances in machinery, but adjustments made to human acceptance and understanding of ourselves and of the place that robots will have in our lives. This chapter is concerned with what shapes our assessment of a robotic counterpart in the context of collaboration, and how the future of robotics is affected by these determinants.

The definition of a social robot has been examined in depth [14, 35, 25, 11], however the term has progressed from its original meaning inspired by the collective behavior of insects, to an association more closely aligned with 'anthropomorphic social behavior' [14]. Cynthia Breazeal offers further classification of social robots in accordance with the complexity of the interaction supported, however for the broader purposes of this paper we align ourselves with Terrence Fong's definition of 'socially interactive robots' in which social interaction plays a key role [30]. Fong's definition refers to a specific type of social robots, who act as partners, peers or assistants, and he distinguishes these robots from other robots that involve 'conventional' human-robot interaction such as teleoperation scenarios. We adopt this definition because Fong's 'socially interactive robots' specifically require considerations for the human in the loop, as an interactive partner [30].

In looking at a variety of social robots presented in the last fifteen years, including [31, 58, 46, 22, 18, 37, 63, 49, 36, 57, 28, 45], we have identified that although these robots varied vastly in function and application, a common trajectory of motivation for the authors emerged. Deeper interaction, collaboration, and more subtle, sustainable and meaningful human-robot relationships were cited as the driving motivation behind their work. The authors sought 'increasing acceptance and integration into people's lives' [45], 'longer and ... a much more rewarding interaction' [37], and 'to combine human creativity, emotion, and aesthetic judgment with the algorithmic computational capabilities of computers, allowing human and artificial players to build on each other's ideas' [63]. We acknowledge the focused nature of our sample, however the sample was curated across fifteen years of robots shown at non-traditionally aligned robotics conferences such as ACM SIGGRAPH and ACM CHI. These robots were specifically selected as breaking from the traditional classifications of industrial, service and social machines, and functioned more like cultural agents that were performative or creators of material and nonmaterial culture, and therefore might be more aptly classified as 'Cultural Robots'. This discussion is expanded in the following sections.

The collective desire of these contemporary roboticists for deeper interaction and greater collaboration extends beyond the mechanics of the robot to inherently include working with others, namely, a human counterpart. For the purposes of this study, we will focus on the desire for collaboration, as by nature it encompasses the other aforementioned qualities. As outlined by Breazeal [17], collaboration demands an open channel of communication, maintenance of shared goals and beliefs amongst team members, mutual understanding and joint activity, and established roles and capabilities amongst team members [20]. We envisage human-robot collaborative tasks to take place in domestic, institutional and commercial settings, including but not limited to, household duties (cooking, cleaning etc.), cultural undertakings such as composing a piece music, or painting a picture, as well as day-to-day operations such as caring for the elderly, or building a piece of furniture.

While extensive research has been conducted concerning the technical and cognitive demands on interactive and collaborative robots [2, 9, 15], as highlighted by Fong [30], little research has been done to investigate human willingness to closely interact with a social robot. Brian Duffy confirms this need, 'The issue of social acceptance is important in the greater debate regarding the context and degree of social integration of robots into people's social and physical space [24].' To that end, we have developed a cognitive model adapted from the Theory of Planned behavior, to investigate the influences and determinants of human willingness to collaborate with a robot.

7.2 A COGNITIVE MODEL FOR DETERMINING HUMAN WILLINGNESS TO ENGAGE WITH COLLABORATIVE ROBOTS

The cognitive model developed (Figure 7.1) is an adaptation of the Theory of Planned Behavior (TPB) [4], appropriated from the field of psychology.

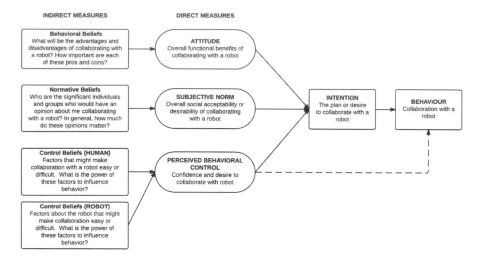

FIGURE 7.1   A cognitive model for human willingness to collaborate with a robot based on the Theory of Planned behavior.

TPB is 'a social cognitive model of the influences on an individual's decision to engage in a particular behavior, (and) has been identified as a promising framework [19] for detecting the known physical, psychological and social determinants (...) most strongly impacting expectations [26]'. The TPB is designed to predict and explain human behavior in particular circumstances [3], on the premise that a person's behavioral intentions are the most proximal and strongest determinant of human behavior [26].

The TPB proposes three primary determinants of intention. The first is the attitude towards the behavior, which refers to the degree to which a person has an overall positive or negative appraisal of the behaviour, in this case, collaborating with a social robot, and how important this appraisal is (behavioral beliefs). The second is labeled the subjective norm, and refers to the perceived social pressure to perform or not perform the behavior. Inclusive to this is the normative beliefs- an assessment of the overall importance of the opinions of significant others. The third determinant is perceived behavioral control, which is the perceived ease or difficulty of performing the behavior, based on an assessment of the self and all available resources, and the power of this belief to influence the individual's behavior (control beliefs) [3].

Many different models have emerged concerning the adoption of technology [48], drawing from information technology, psychology, and the social sciences. Some of these models include Technology Acceptance Models [21], which in this case is also a derivative of the Fishbein model, and looks to broadly measure a user's overall attitude towards a technological system. Diffusion of Innovation models are used to analyse the communication and adoption of new ideas into society, and to combat consumer uncertainty [52]. Technology, Organization and Environment frameworks identify three aspects of an enterprise's context that influence the process by which it adopts and implements a technological innovation: technological context, organizational context, and environmental context [61]. While these models can assist us in understanding some environmental factors, adoption of technology trends and perhaps individual readiness to engage with technology [48], without the additional determinants given by Icek Ajzen, such as the Perceived Behavioral Control, they are not nearly as comprehensive in understanding intended behaviour with limited variables, and are not assessing personal social and cognitive determinants. Furthermore, the TPB has been given rigorous examination and application in literature and research, and has found to be a strong predictor of intention and behavior over a variety of applications [3]. Other diverse applications include predicting healthy eating by students [12]; Influenza vaccination [23]; and willingness to pay for green freight transportation [54].

It should be noted that when used in psychology, the Theory of Planned Behavior has the potential to be operationalized through the development of a questionnaire using standard procedures described by [4]. However for the purposes of this application we are employing the model more broadly to outline the relevant constructs determining human behavior.

Limitations of the model have been raised by critics [55, 33] and acknowledged by its author [4]. Criticisms include the vulnerable nature of self-reporting [7], and that the prediction of behavior from intentions can involve potential problems (for detailed discussion see [29]). Beliefs that are accessible in a hypothetical situation (i.e. during self-reporting) may also differ greatly from those that occur in a real situation in which the behavior is occurring [6]. However, in the case of collaboration with a robot (and indeed any new technology), changes to attitudes are often beneficial, as initial apprehension may be overcome during prolonged exposure to the technology, 'people interacting with a product may form an overall assessment within a fraction of a second, but this assessment may change as they process additional information' [27].

In Figure 1, the original TPB model has been modified with an additional predictor delineating human control beliefs about a robot. Ajzen (author of the TPB) encourages the addition of new predictors to the TPB, adding that they can significantly improve the prediction of intention, as the new measure may contain meaningful variance in attitudes not accounted for in that particular scenario by the original model [4]. In predicting human willingness to collaborate with a robot, given the aforementioned variety of social robots in both function and aesthetics, we hypothesize that the user's assessment of different robots in different scenarios would significantly impact their intention to collaborate. Thus, we have separated the control beliefs concerning the assessment of the robot, to both indicate its significance, and to account for the variance in willingness to collaborate as a response to individual robots, and not necessarily to the self.

In his meta-analysis of the TPB, Christopher. J. Armitage explains that the Perceived Behavioural Control is found to have direct influence, and could account for significant variance in both intention and behavior, and could be independent of the two other constructs [8]. For this reason, we have chosen to focus our discussion in this chapter on the Perceived Behavioural Control, in particular, the control beliefs concerning a social robot. A brief discussion of the Behavioural and Normative Beliefs however further establishes the significance of culture in the assessment of willingness to collaborate with a robot.

A number of cognitive models have been formulated concerning cognition in social robots [39], however to the authors' knowledge, no cognitive model or comprehensive study exists specifically examining human cognition in willingness to collaborate with a robot.

7.2.1 Determinants

7.2.1.1 Control Beliefs

In focusing on Perceived Behavioral Controls, we have separated our analysis into two predictors of Control Beliefs, (HUMAN) and (ROBOT). PBC is the perceived ease or difficulty of engaging in the behavior determined by belief about personal resources, knowledge, skills and opportunities, and the power of these elements to influence behavioral outcomes [26].

In a scenario requiring collaboration, the collaborators must establish mutual beliefs determining the intention, values, motivation and the capabilities of one another [17]. Thus, the human collaborator must make this assessment of a robotic counterpart.

The 'media equation' presented by Byron Reeves and Clifford Nass [51] tells us that people will equate social robots with real social actors, and that our interactions with new media are social and natural, just like real life. Nass and associates have also found that individuals automatically apply social rules in their interaction with computers as if they were interacting with real human beings [42, 47]. Considering this, we can assume that a person will assess a robot collaborator just as they might assess a human. For a robot collaborator, this judgement may be a matter of cognitive wiring, but for the human counterpart, we argue that this assessment will primarily be an initial visual aesthetic analysis. This analysis will be governed by inherent cultural expectations and readings. Fong explains, 'The form and structure of a robot is important because it helps establish social expectations. Physical appearance biases interaction. Moreover, the relative familiarity (or strangeness) of a robot's morphology can have profound effects on its accessibility, desirability, and expressiveness [30].

Frank Hegel discusses that a robot's physical appearance is the most visible and easily accessible information about its capabilities [34]. This is corroborated by Jennifer Goetz, 'a robot's appearance and behavior provide clues that influence perceptions of the robot's propensities and assumptions about its capabilities [32]. Furthermore, Goetz summarises Aarts in saying, 'Psychology tells us that people's initial response to a robot will be fast, automatic (unconscious) and heavily stimulus or cue-driven' [1]. This fast, automatic response to a robot's appearance tells us that the assessment of a robot that will shape a person's willingness to collaborate, at least in the short term, will be primarily based upon its visual aesthetic features, before any interaction has begun.

Micki Einseman affirms that visible design attributes communicate not only function and capabilities, but also aesthetic and symbolic information [27]. He explains that symbolic information pertains to meanings and associations to a product beyond its basic utility. In the same way, a robot's appearance may have the faculty to communicate not only its function, but also more subtle idiosyncratic information, such as values, desires, culture or intention.

Perceived behavioural control (PBC) is assumed to reflect past experiences as well as anticipated impediments or obstacles [4] therefore we must account for the impact of predispositions, fears or assumptions made about robots that have been established either through experience with other robots or exposure to popular culture such as science fiction writings and other media. Zayera Khan explains that the influence of science fiction can fuel both fear and fascination with robots [38].

7.2.1.2 Behavioral Beliefs

Behavioural Beliefs pertain to an individual's global positive or negative evaluation and attitude towards performing a particular behaviour [8]. Beliefs about perceived benefits have been shown to be a significant determinant across a variety of fields, such as the positive influence of perceived health benefits in relation to the consumption of organic food [64]. Beliefs concerning the perceived benefits or detriments of collaboration with a robot will be dependant on various individual situations, reasons for the collaboration, and the different individuals and robots involved.

7.2.1.3 Normative Beliefs

Normative Beliefs describe an individual's beliefs about what others expect them to do in a given situation, and are in turn multiplied by their motivation to comply with these perceived norms [5]. Cross-cultural studies have demonstrated that Normative Beliefs have a strong capacity to influence an individual's intention to perform a behaviour, even in a 'Confucian culture' (p.193) where collective and interdependent mentality may be more highly valued, compared to Western cultures [40]. Given the highly social and primal nature of Normative Beliefs and their global application, further research into their specific influence upon an individual's willingness to collaborate with a robot is necessary and potentially very beneficial, but beyond the scope of this chapter.

Considering the salience of socio-cultural influences in determining human willingness to engage with robotic agents, combined with the discernible emergence of robots as participants in and producers of culture, we propose the following definition of a potential area of research that we have defined as Cultural Robotics.

7.3 CULTURAL ROBOTICS

We define Cultural Robotics tentatively as *the study of robots in culture, cultural acceptance of robots, robot-cultural interaction and robot-generated cultures.* A Cultural Robot can be loosely defined as *an autonomous robotic entity that contributes to the development of material and/or nonmaterial culture.*

We acknowledge the notion of 'culture' as having an array of definitions across a number of disciplines. Overall, it has emerged that culture is a multilayered construct, inclusive of not only external artefacts such as language and customs, food and dance, but also nuanced elements of 'a group's shared set of specific basic beliefs, values, practices and artefacts that are formed and retained over a long period of time' [59].

We have identified culture as an important factor to consider in the design of a social robot for a number of key reasons. Many social roboticists describe the need for robots to acquire 'social' skills and awareness, under the assumption that these norms are globally applicable. While skills surrounding basic interaction may be so, much of human interaction is shaped by specific cultures, for example, styles of greeting, personal space [43], volume and pace of speech, manners, formalities, eating and drinking rituals, etc.

In addition to this, many of the qualities required to be shared by both parties in collaboration are dependant on or determined by culture, such as shared values, beliefs, goals and priorities. As such, a social collaborative robot will need to be built in consideration of the specific culture it will become a part of or be collaborating with, and the way in which it may or may not reflect this culture through its physical appearance and operation; 'Socially interactive robots will eventually need to support a wide range of users: different genders, different cultural and social backgrounds [30].'

In the field of computing, Matthias Rauterberg expands on common understandings of culture [44], to explain the deeper applications of culture to technology and interaction, 'Cultural Computing is more than integrating cultural aspects into the interaction. It is about allowing the user to experience an interaction that is closely related to the core aspects of his/her culture [50].' The integration of robots into culture and culture into robots is proposed by [53], who propose a bidirectional shaping of society in which technology and society shape one another.

When we talk about considering culture when designing a robot, we mean a number of things. Firstly, a robot as an actor in a particular culture, which (we would argue), all social robots are. Hooman Samani et al. ([53]) quotes Lucy Suchman who argues that users interact with technology using similar expectations as in human-to-human communication, and in doing so, they shape the properties of robots using their cultural and social values [56]. Suchman implies that robots are not only built reflective of a particular culture, but are continually shaped and analysed through the lens of each person's individual culture, as they see and interact with it.

The second stage of cultural integration is a robot as a participant in, or producer of human material and nonmaterial culture. This is the level of integration where we would position most current social robots. Simply by interacting with people, robots are exposed to and influenced by nonmaterial culture, but through collaboration, social robots can now also help cook a meal [58], improvise on a piece of music [63], or serve a drink [31]. As this stage progresses, we may see robots begin to independently produce cultural artefacts, and as this grows, we see the potential for the advent of robotic community culture.

Robotic community culture, as first outlined by Samani et al. in [53], is the emergence of artificial culture in robotic communities, and refers to the creation of artefacts, values, beliefs, customs and other cultural dimensions by robotic agents. This stage of robotic culture has not yet been actualized, but is not beyond the foreseeable future. Samani et al. assert that through stages such as local embodied robotic interactions, norm modelling, self-replication and robotic co-operation and coherent behaviour, robotic cultural norms have the potential to emerge. However, as aptly identified by Susan Blackmore, these manifestations of machine culture may be completely inalienable or inscrutable to human beings [13]. The discussion of the emergence of autonomous robotic culture is beyond the scope of this paper, however, it is important to keep in mind as the 'bidirectional shaping' of technology and culture continues.

In considering culture, we must also acknowledge that not all cultures are receptive to robots in the same way, and in fact exhibit strong and particular preferences when it comes to the type of robot, and what function it will have in their lives. Some early research has been done comparing the culturally variable preferences for robotic design including [41, 10]. Different cultures (in this case, countries) varied greatly on accounts of anthropomorphism, intelligence, robotic form, interactivity and general attitudes towards robotics. Both of these studies were conducted with static images of robots alone, and both studies involved less than 100 participants from any given country, who then went on to represent the preferences of their entire nation in the results. These generalization do not give consideration to the vast differences in community or state specific culture, where the views of a Texan farmer may vary greatly from a New York city lawyer. The outcome of these studies may vary with the use of embodied robots, and the variations of cultural preferences within a country would make for an interesting study.

However, similar to the results of these studies, Reeves and Nass found that people had responses to extremely minimal cues about a computers' 'personality'. Computer name, voice, confidence, and sequencing of action were all interpreted as cues to the personality and background of the computer, and led to people liking and disliking, associating or disassociating with it [51]. These results verify the vast and complicated implications of culture for robotic design.

In returning to our examples of [58, 63, 31], we see that contemporary robotics development presented at forums outside traditional robotics conferences such as IROS, ICRA and RO-MAN show that a new breed of roboticist is engaging with applications that lie outside of the traditional industrial, service and social/therapy applications. These robots are explicitly made to interact with humans in a cultural capacity, whether it is inheriting and appropriating the current behaviours and mannerisms of a stereotypical bartender, to a cooking partner technology that participated in the preparation of cuisine, these robots are presented in the guise of social robots, but their purpose and functionality lean more towards that of active cultural agents, and we see more and more of these culturally specific social and HRI-motivated robots emerge every day. With this in mind, coupled with cultural determinants proving significant in light of the cognitive model presented in this paper, we foresee the emergence of a new field of robotics research and classification of robotic agents that can be best described as Cultural Robotics.

7.4 FUTURE RESEARCH

The results of this early research have provoked a number of significant questions and highlighted areas for future research.

Through the implementation of user studies and design iteration, we have begun work to test the cognitive model presented to determine the extent to which the determinants are able to predict human willingness to collaborate with a robot, and to investigate alternative possibilities for how it may be operationalised. Furthermore, we are interested in how these determinants might be distilled into a framework for use in the design and planning of socio-cultural robotic agents.

We are curious to determine whether this research will result in the identification and deliberate integration of cultural semiotics into robotic design, or encourage an awareness and reduction of culture in design, leading to the development of 'culturally neutral' robots, or possibly both.

This cognitive model requires further research into specifically what factors may influence an individuals assessment of their own capacity and resources to engage with a robot (see HUMAN Control Beliefs). We predict that some of these factors may be deduced as a result of aesthetics, for example a complicated mechanical exterior compared to a streamline and simplistic aesthetic may greatly alter not only the user's assessment of the robot but also their personal capacity to engage with it.

Finally, we would like to further establish and refine the definition of Cultural Robotics through engagement with and feedback from researchers and roboticists in a variety of fields.

7.5 CONCLUSION

In this chapter we discuss the determinants of human willingness to collaborate with robots. Based on the previous writings of researchers such as Fong, Breazeal, Duffy and others, as well as analysis of contemporary robotic creations such as [58, 63, 31], we surmise that a new type of robot is emerging on the fringe of social robotics that could be classified as 'cultural' robots. Furthermore, the meta-motivation for the development of these new types of robots is a collective desire amongst roboticists for deeper interaction and greater collaboration between robot and human counterparts. Using these same robots as well as other examples, we outline how future robots will increasingly become enabled with cultural agency, and that a model for human willingness to collaborate with such robots indicates that cultural considerations are both expanding robotic agency, and necessary for forging more meaningful interactions with humans.

The cognitive model presented in this chapter attempts to outline and understand these determinants and their relation to human-robot interaction. Based on the Theory of Planned Behaviour, we briefly discuss the Behavioural, Normative and Control Beliefs, focusing on the assessment a human actor will make of a robotic collaborative partner, and the influence of aesthetics and culture upon this.

Considering the emergence of a new type of robot, combined with the evident influence of culture upon willingness to engage with social robotic agents, we propose *Cultural Robotics* as a new area of research in robotic development.

7.6 Bibliography

[1] Henk Aarts and Ap Dijksterhuis. Habits as knowledge structures: automaticity in goal-directed behavior. *Journal of personality and social psychology*, 78(1):53, 2000.

[2] Bryan Adams, Cynthia L Breazeal, Rodney Brooks, and Brian Scassellati. Humanoid robots: a new kind of tool. Technical report, DTIC Document, 2000.

[3] Icek Ajzen. The theory of planned behavior. *Organizational behavior and human decision processes*, 50(2):179–211, 1991.

[4] Icek Ajzen. The theory of planned behavior is alive and well, and not ready to retire: a commentary on Sniehotta, Presseau, and Araújo-Soares. *Health Psychology Review*, (ahead-of-print):1–7, 2014.

[5] Icek Ajzen and Martin Fishbein. Attitudes and normative beliefs as factors influencing behavioral intentions. *Journal of Personality and Social Psychology*, 21(1):1, 1972.

[6] Icek Ajzen and James Sexton. Depth of processing, belief congruence, and attitude-behavior correspondence. *Dual-process theories in social psychology*, The Guilfield Press, New York, NY, pages 117–138, 1999.

[7] Christopher J Armitage and Mark Conner. Social cognition models and health behavior: A structured review. *Psychology and Health*, 15(2):173–189, 2000.

[8] Christopher J Armitage and Mark Conner. Efficacy of the theory of planned behavior: A meta-analytic review. *British journal of social psychology*, 40(4):471–499, 2001.

[9] Tucker Balch and Ronald C Arkin. Communication in reactive multia-gent robotic systems. *Autonomous Robots*, 1(1):27–52, 1994.

[10] Christoph Bartneck. Who like androids more: Japanese or us Americans? In *Robot and Human Interactive Communication, 2008. RO-MAN 2008. The 17th IEEE International Symposium on circuits and systems*, pages 553–557. IEEE, 2008.

[11] Christoph Bartneck and Jodi Forlizzi. A design-centred framework for social human-robot interaction. In *Proceedings of the 13th IEEE International Workshop on Robot and Human Interactive Communication, Kurashiki*, pages 591–594, 2004.

[12] Evagelos Bebetsos, Stiliani Chroni, and Yannis Theodorakis. Physically active students'intentions and self-efficacy towards healthy eating. *Psychological Reports*, 91(2):485–495, 2002.

[13] Susan Blackmore. Consciousness in meme machines. *Journal of Consciousness Studies*, 10(4-5):4–5, 2003.

[14] Cynthia Breazeal. Toward sociable robots. *Robotics and autonomous systems*, 42(3):167–175, 2003.

[15] Cynthia Breazeal and Brian Scassellati. How to build robots that make friends and influence people. In *Intelligent Robots and Systems, 1999. IROS'99. Proceedings. 1999 IEEE/RSJ International Conference on Intelligent Robotics and Systems*, volume 2, pages 858–863. IEEE, 1999.

[16] Cynthia L Breazeal. *Sociable machines: Expressive social exchange between humans and robots*. PhD thesis, Massachusetts Institute of Technology, 2000.

[17] Cynthia L Breazeal. *Designing sociable robots*. MIT press, Cambridge, MA, 2004.

[18] Andrew G Brooks, Jesse Gray, Guy Hoffman, Andrea Lockerd, Hans Lee, and Cynthia Breazeal. Robot's play: interactive games with sociable machines. *Computers in Entertainment (CIE)*, 2(3):10–10, 2004.

[19] Sandra Brouwer, Boudien Krol, Michiel F Reneman, Ute Bültmann, Renée-Louise Franche, Jac JL van der Klink, and Johan W Groothoff. Behavioral determinants as predictors of return to work after long-term sickness absence: an application of the theory of planned behavior. *Journal of occupational rehabilitation*, 19(2):166–174, 2009.

[20] Philip R Cohen, Jerry L Morgan, and Martha E Pollack *Intentions in communication*. MIT press, Cambridge, MA, 1990.

[21] Fred D Davis. *A technology acceptance model for empirically testing new enduser information systems: Theory and results*. PhD thesis, Massachusetts Institute of Technology, Cambridge, MA, 1985.

[22] Carl Di Salvo, Francine Gemperle, Jodi Forlizzi, and Elliott Montgomery. The hug: an exploration of robotic form for intimate communication. In *Robot and Human Interactive Communication, 2003. Proceedings. ROMAN 2003. The 12th IEEE International Workshop on Robot and Human Interactive Communication*, pages 403–408. IEEE, 2003.

[23] SF Doris, Lisa PL Low, Iris FK Lee, Diana TF Lee, and Wai Man Ng. Predicting influenza vaccination intent among at-risk chinese older adults in hong kong. *Nursing research*, 63(4):270–277, 2014.

[24] Brian R Duffy. Anthropomorphism and the social robot. *Robotics and autonomous systems*, 42(3):177–190, 2003.

[25] Brian R Duffy, Colm Rooney, Greg MP O'Hare, and Ruadhan O'Donoghue. The Social Robot. In *Robotics and Autonomous Systems*. 2000.

[26] Debra A Dunstan, Tanya Covic, and Graham A Tyson. What leads to the expectation to return to work? insights from a theory of planned behavior (tpb) model of future work outcomes. *Work: A Journal of Prevention, Assessment and Rehabilitation*, 46(1):25–37, 2013.

[27] Micki Eisenman. Understanding aesthetic innovation in the context of technological evolution. *Academy of Management Review*, 38(3):332–351, 2013.

[28] Charith Lasantha Fernando, Masahiro Furukawa, Tadatoshi Kurogi, Kyo Hirota, Sho Kamuro, Katsunari Sato, Kouta Minamizawa, and Susumu Tachi. Telesar v: Telexistence surrogate anthropomorphic robot. In *ACM SIGGRAPH 2012 Emerging Technologies*, page 23. ACM, 2012.

[29] Martin Fishbein and Icek Ajzen. *Predicting and changing behavior: The reasoned action approach*. Taylor & Francis, New York, NY, 2011.

[30] Terrence Fong, Illah Nourbakhsh, and Kerstin Dautenhahn. A survey of socially interactive robots. *Robotics and autonomous systems*, 42(3):143–166, 2003.

[31] Mary Ellen Foster, Andre Gaschler, and Manuel Giuliani. How can I help you': comparing engagement classification strategies for a robot bartender. In *Proceedings of the 15th ACM on International Conference on Multimodal Interaction*, pages 255–262. ACM, 2013.

[32] Barbara J Grosz. Collaborative systems (AAAI-94 presidential address). *AI magazine*, 17(2):67, 1996.

[33] Wendy Hardeman, Marie Johnston, Derek Johnston, Debbie Bonetti, Nicholas Wareham, and Ann Louise Kinmonth. Application of the theory of planned behavior in behavior change interventions: A systematic review. *Psychology and health*, 17(2):123–158, 2002.

[34] Frank Hegel. Effects of a robot's aesthetic design on the attribution of social capabilities. In *RO-MAN, 2012 IEEE*, pages 469–475. IEEE, 2012.

[35] Frank Hegel, Claudia Muhl, Britta Wrede, Martina Hielscher-Fastabend, and Gerhard Sagerer. Understanding social robots. In *Advances in Computer-Human Interactions, 2009. ACHI'09. Second International Conferences on Advances in Computer-Human Interaction*, pages 169–174. IEEE, 2009.

[36] Interbots.com,. Interbots :: We Build Character. N.p., 2015. Web. 20 Dec. 2014.

[37] Mattias Jacobsson, Sara Ljungblad, Johan Bodin, Jeffrey Knurek, and Lars Erik Holmquist. Glowbots: robots that evolve relationships. In *ACM SIGGRAPH 2007 emerging technologies*, page 7. ACM, 2007.

[38] Zayera Khan. Attitudes towards intelligent service robots. *NADA KTH, Stockholm*, 17, 1998.

[39] Stefan Kopp and Jochen J Steil. Special corner on cognitive robotics *Cognitive processing*, 12(4):317–318, 2011.

[40] Chol Lee and Robert T Green. Cross-cultural examination of the fishbein behavioral intentions model. *Journal of international business studies*, pages 289–305, 1991.

[41] Hee Rin Lee and Selma Sabanović. Culturally variable preferences for robot design and use in South Korea, Turkey, and the United States. In *Proceedings of the 2014 ACM/IEEE International Conference on Human-Robot Interaction*, pages 17–24. ACM, 2014.

[42] Kwan Min Lee, Wei Peng, Seung-A Jin, and Chang Yan. Can robots manifest personality?: An empirical test of personality recognition, social responses, and social presence in human–robot interaction. *Journal of communication*, 56(4):754–772, 2006.

[43] Jacob Lomranz. Cultural variations in personal space. *The Journal of Social Psychology*, 99(1):21–27, 1976.

[44] David Matsumoto and Linda Juang. *Culture and psychology*. Cengage Learning, Wadsworth, USA, 2013.

[45] Sean McGlynn, Braeden Snook, Shawn Kemple, Tracy L Mitzner, and Wendy A Rogers. Therapeutic robots for older adults: investigating the potential of Paro. In *Proceedings of the 2014 ACM/IEEE International Conference on Human-Robot Interaction*, pages 246–247. ACM, 2014.

[46] Yasushi Nakauchi and Reid Simmons. A social robot that stands in line. *Autonomous Robots*, 12(3):313–324, 2002.

[47] Clifford Nass and Youngme Moon. Machines and mindlessness: Social responses to computers. *Journal of social issues*, 56(1):81–103, 2000.

[48] Tiago Oliveira and Maria Fraga Martins. Literature review of information technology adoption models at firm level. *The Electronic Journal Information Systems Evaluation*, 14(1):110–121, 2011.

[49] Parorobots.com,. PARO Therapeutic Robot. N.p., 2015. Web. 20 Dec. 2014.

[50] Matthias Rauterberg. From personal to cultural computing: how to assess a cultural experience. *uDayIV–Information nutzbar machen*, pages 13–21, 2006.

[51] Byron Reeves and Clifford Nass. The media equation. 1996. *CSLI and Cambridge, Cambridge*.

[52] Everett M Rogers. *Diffusion of innovations*. Simon and Schuster, New York, NY, 2010.

[53] Hooman Samani, Elham Saadatian, Natalie Pang, Doros Polydorou, Owen Noel Newton Fernando, Ryohei Nakatsu, and Jeffrey Tzu Kwan Valino Koh. Cultural robotics: The culture of robotics and robotics in culture, In *International Journal of Advanced Robotic Systems*, 2013.

[54] Dara G Schniederjans and Christopher M Starkey. Intention and willingness to pay for green freight transportation: an empirical examination. *Transportation Research Part D: Transport and Environment*, 31:116–125, 2014.

[55] Falko F Sniehotta, Justin Presseau, and Vera Araújo-Soares. Time to retire the theory of planned behavior. *Health Psychology Review*, 8(1):1–7, 2014.

[56] Lucy Suchman. *Human-machine reconfigurations: Plans and situated actions.* Cambridge University Press, 2007.

[57] Yuta Sugiura, Takeo Igarashi, Hiroki Takahashi, Tabare Akim Gowon, Charith Lasantha Fernando, Maki Sugimoto, and Masahiko Inami. Graphical instruction for a garment folding robot. In *ACM SIGGRAPH 2009 Emerging Technologies*, page 12. ACM, 2009.

[58] Yuta Sugiura, Anusha Withana, Teruki Shinohara, Masayasu Ogata, Daisuke Sakamoto, Masahiko Inami, and Takeo Igarashi. Cooky: a cooperative cooking robot system. In *SIGGRAPH Asia 2011 Emerging Technologies*, page 17. ACM, 2011.

[59] Vas Taras, Julie Rowney, and Piers Steel. Half a century of measuring culture: Review of approaches, challenges, and limitations based on the analysis of 121 instruments for quantifying culture. *Journal of International Management*, 15(4):357–373, 2009.

[60] Sebastian Thrun, Michael Beetz, Maren Bennewitz, Wolfram Burgard, Armin B Cremers, Frank Dellaert, Dieter Fox, Dirk Haehnel, Chuck Rosenberg, Nicholas Roy, et al. Probabilistic algorithms and the interactive museum tour-guide robot Minerva. *The International Journal of Robotics Research*, 19(11):972–999, 2000.

[61] Louis G Tornatzky, John D Eveland, and Mitchell Fleischer. Technological innovation: definitions and perspectives. *The Processes of Technological Innovation*, pages 9–25, 1990.

[62] Sherry Turkle. An ascent robotics culture: New complicities for companionship. In *American Association for Artificial Intelligence AAAI*, MIT, Cambridge, MA, 2006.

[63] Gil Weinberg, Guy Hoffman, Ryan Nikolaidis, and Roberto Aim. Shimon+ zoozbeat: an improvising robot musician you can jam with. In *ACM SIGGRAPH ASIA 2009 Art Gallery & Emerging Technologies: Adaptation*, pages 84–84. ACM, 2009.

[64] Lukas Zagata. Consumers? Beliefs and behavioral intentions towards organic food. Evidence from the Czech Republic. *Appetite*, 59(1):81–89, 2012.

[65] Terry, D. J. Self-efficacy expectancies and the theory of reasoned action. *The theory of reasoned action: Its application to AIDS-preventive behaviour. International series in experimental social psychology* 28 (1993): 135–151.

CHAPTER 8

Social Cognition of Robots during Interacting with Humans

Alex Yu-Hung Chien

Department of Computer Science,
National Tsing Hua University, Taiwan.

Von-Wun Soo

Department of Computer Science,
Naitonal Tsing Hua University, Taiwan.

CONTENTS

F OR robots to understand the social context and generate the proper responses is deemed to be a challenging task. Most social interactions
in particular in dialogue among humans require substantial domain and
context-specific knowledge to interpret the intents, emotions and social relations among the agents involved. Lack of the kind of knowledge and models,
it is hard for a robot or virtual agent to infer the social contexts and generate
proper responses.

We established the technologies based on speech-act theories that enable a
virtual agent or a robot to interpret the social contexts in a dialogue. We show
how different speech acts can be modeled and used to assist virtual agents
in inferring the social contexts. To deal with the sequential changes of social
contexts in a dialogue, we adopt dynamic belief networks to compute the
likelihood of the social context parameter values. We evaluate the feasibility
of the social awareness model against human annotated dialogue corpus
excerpted from segments of movie scripts.

8.1   INTRODUCTION

To make robots to behave properly in human society is an ultimate and
challenging goal of robotics research. The major factor for hindering robots
to co-inhabit with humans in human society is lack of social context common
sense which we call it as social context awareness. Without social context
awareness, robots cannot interpret human's intentions and generate proper
response accordingly. By social context wareness, we extend the concept of
current state of the arts of *context awareness* that is to infer the physical and
environmental contexts from sensor data [1] to much higher level of social
contexts in human interactions.

In this chapter, we treat robots as autonomous software agents who will
be able to infer the belief, desire, goal and intention, emotion states, mood,
personality of other agents in a given social interactions. It also means agents
will understand the social relations, roles, commitment of other agents in the
social interactions. Therefore, the terms robots and agents can be used indistinguishable in this paper. To some extent, wider scope of social awareness
also implies agents understand the social norms, social convention, morality

of behaviors or culture of human society. However, without losing of generality, we limit our study on some lower level of social contexts.To interpret a given unknown social context, we might have to rely on other known social contexts to infer correctly. For example, we want to understand a person's intention in conducting certain behavior, we might have to know the person's role or relation with other agents in the social context to assess if it obeys the social norm or not. Since social contexts might change during social interactions, the history social contexts might become important background information to interpret current social context.

For virtual agents or robots to interact and communicate among each other freely, there are a lot of bottlenecks to overcome. Previous virtual agents research assume Grice maxims of communication [12] to simplify the agent communication problem to overcome some bottlenecks that lead to the design of ACL [2] [11][10]. The Grice maxims can be viewed as the assumptions on an ideal communication that must obey the principles of Quality, Quantity, Relation and Manner. The quality principle assumes the communication agents are honest and truthfully. For example, if they ask a question about X, they must don't know X and they really want to know X. The quantity principle assumes the communication agents be precise. For example, if they communicate, they will reveal as much information as is required, no more and no less. The relation principle, assume the communication agents will convey only relevant information. That is, if they refer the things, they must be relevant to the purpose of communication. The manner principle assumes the communication agents must present appropriately. Namely, they will avoid ambiguities and present things in a clear and orderly way. The Grice maxims are nice to characterize an ideal communication among virtual agents and robots. They can also serve as criteria to evaluate whether a way of communication is good. However, it is hard for human to follow in the real world communication even human communicators consciously might wish to attain the ideal communication goals. There are couples of reasons that human communicators cannot follow the Grice maxims easily:

1. Humans are not always cooperative, and therefore, you cannot assume human communicators answer the question based on what they know. They might not be polite and therefore do not always follow the conventional protocol in communication or even worse they will sometimes violate the norms of communications in a hostile manner. For example, you are asking a stranger some information, the stranger might not necessarily reply what they know to you.

2. Humans are not always certain about what they know or cannot sometimes find the exact terms or use parsimony statements to express. Or they might sometimes have false beliefs, even they have an intention to reveal the truth. Therefore they can sometimes generate inexact or false information unintentionally.

3. Humans might not always be honest, and can sometimes lie or cheat to others. Therefore they might also generate false information intentionally.

4. Humans might be malicious or negative pragmatic such as irony or sarcasm at various situations in social interactions. For example, one might want to discredit or hurt the social influence of another.

5. Human might not express directly, but rather adopt indirect hints (sometimes due to politeness) in social interactions. For example, not to explicitly mention the age or sex to female persons directly.

6. Humans may have ambiguous or multiple intentions in one statement. For example, asking time or requesting to light a cigarette from a person might be sometimes used as a *start of conversation* to a stranger but might also have an intention to request the time or assistance at the same time.

Social contexts in human interactions, either physical or dialogical, can be inferred from various clues based on observations.The fundamental techniques to interpret the social contexts can be based on various sensor technologies for context-awareness [9] such as camera, sound, or even environmental sensors such as GPS, temperature, etc. However, social context awareness involves many mental states of agents that is difficult to verify or directly observable. With sensor data alone, we cannot easily distinguish social mental states of agents without support from some context background information and social models. This is because without background social contexts, there can be many ambiguities and exceptions in the interpretation. For example, seeing someone waving a hand, it might mean saying hello, saying good bye, or saying no, etc. depending on different social situations. Without social background context models, only limited levels of abstract inference can be made based on sensor data alone. For example, camera sensor data might be able to recognize a person's body movement as *walking*, but we must combine with background contexts such as walking toward/out of a room at given time period to infer if the person is going to sleep or is going out to work, etc. Sometimes we even need to know the social role of a person to judge if his/her behavior is permitted to enter the room.

In reality, both physical interactions and verbal interactions can be viewed as communication actions. Agents can use body language, gesture, facial expressions etc to conduct communication among each other. But verbal interaction via dialogue is more complex that requires agents to adopt various complex levels of natural language processing and reasoning techniques. We attempt to capture the knowledge of verbal interactions in terms of speech act theories and models [2]. The speech act model is to treat each utterance in a dialogue as an action that has intention, preconditions and effects. The pragmatic knowledge of a speech act is expressed in terms of the change of preconditions and effects. In particular, we model a speech act in terms of the change of social contexts in both preconditions and effects of the speech act.

Since the social context states are usually not directly observable as agent's mental states, we attach a probability to express the strength of the influence of the change of social context states of the speech act.

In the following sections we will discuss the difference between speech acts against physical act and explain how dialogue contexts can be expressed in terms of agent mental states in section 8.2, and our proposal of modeling speech act as diaolgue context change in secion 8.3. In section 8.4, we elaborate the dialogue context as physical context, emotion model, social relation, social role, personality, and preference and show in section 8.5 how a speech act model can be expressed in terms of these dialogue contexts with probabilities attached to conditions. In section 8.6, we the sequential effects of speech acts in dialogue sequence that will lead us to apply the computation model of Dynamic Bayesian Netowrk (DBN) to conduct context awareness reasoning. Evaluation of our models and methods are explained in section 8.8 from many aspects. First, we show how emotion context of a movie character can be inferred from dialogue sequence in subsection 8.8.2. Second, we show how sensitive a DBN model can cope with noise in speech act input in 8.8.3. Third, we evaluate the feasibility of prediction of pragmatic inference based on semantic speech act sequence with or without social contexts in subsections 8.8.4 and 8.8.5 respectively. In section 8.9, we make conclusion.

8.2   SPEECH ACTS AND DIALOGUE CONTEXT

What is a speech act? Each utterance in a dialogue can be viewed as an action that is similar to a physical action that has preconditions as feasible conditions for the physical action to be carried out and effect conditions as possible outcome states that can be resulted in after the action is completed. However, the speech act are mostly conditioned by the mental states of agents and possible influence the mental states of other agents. The advantages of modeling speech acts to social context awareness is due to that the social context changes in the interactions of speech acts during dialogue can be inferred by the sequence of speech acts. On the other hand, social contexts can also be used to interpret why a speech act is properly chosen to cope with desired change of social context for a speaker agent.

An action can only be applied when all preconditions are matched in the action model. In the other hand, the speech act model is less rigid than the action model. There is no physical restriction for agents to apply speech acts in a conversation, so agents can use any speech acts in any situations. However, dialogues could be unreasonable if agents choose speech acts randomly. In order to make agents understand each other in the conversation, speech acts must be triggered by corresponding dialogue contexts. Although there are many related dialogue contexts that can trigger a speech act, we assume that a speech act can be triggered by a partial match with related dialogue contexts. For example, the emotion context: negative and the institutional relation context: subordinate-to-superior are both related to the speech act

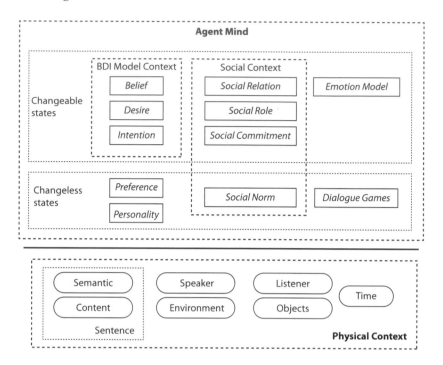

FIGURE 8.1 Contexts in agent mental states vs. a physical context.

beg However, an agent might still *beg* someone for some benefits even if they did not experience the negative emotion. This is a major difference between the speech act model and the action model.

In our speech act model, a speech act affects not only the speaker's mind, but also the listener's mind. All mind contexts of the agents cannot be combined because the mind states of each agent are formulated by independent Dynamic Bayesian Network (DBN) models [15]. This is explained further in 8.7.

In Figure 8.1, we illustrate the flow how an agent uses a speech act to change participators' mind states. First, agents choose a speech act by his/her mind states, such as Belief-desire-intention (BDI) model [17], emotion context [16] and so on. When the speech act is expressed, every participator can receive the speech act and other physical context information from the environment. In the ending phase, each participator's mind states could be changed by the effects of the speech act. Then the next speaker starts to choose a speech act in another cycle.

As discussed above, if we assume that all agents conduct dialogues by obeying Grice's maxims, we could convert the dialogue content semantics directly into pragmatics (infer speech acts directly from their dialogue content semantics). However, in emergent narratives, we forego the assumption of Grice's maxims for agent communication, and therefore, could not directly obtain the pragmatics or speech acts from the dialogue content semantics. Since the agent mind states are essentially unobservable, we cannot confirm whether the pragmatics in a dialogue sequence of inferred speaker agents are indeed the true intentions (or speech acts) of the speakers. Therefore, we distinguish between two different dialogue sequences:

1. a coherent semantic sequence, in which all the dialogue semantics of speech acts in the dialogue sequence can be treated as pragmatics and no conflict in dialogue contexts of those speech acts can be found in the sequence; and

2. an incoherent semantic sequence in which the dialogue semantics of speech acts are treated directly as pragmatics, and if so, the dialogue contexts of certain speech acts might be in conflict with their true pragmatics.

Dialogues in the emergent narratives are generally incoherent semantic sequences, and therefore, we must find an explanation of the most likely pragmatic speech act sequence for a given dialogue sequence. To achieve this, we should make agent aware the dialogue contexts in the agent conversation.

8.3 MODEL SPEECH ACTS AS DIALOGUE CONTEXT CHANGE

All physical actions must obey physical laws to be executable. A physical action usually can only be applied when preconditions in the action model are matched. In contrast, the speech act model is less rigid than the physical action model, because, agents can carry out any speech act at arbitrary situations without restrictions. However, if dialogues are conducted in an arbitrarily manner, they could become odd and not be easily understandable. Therefore, in order to make agents understandable among each other in dialogical interaction, speech acts must be triggered or applied at proper dialogue contexts even they do not have very rigid constraints. We thus assume that a speech act can be triggered by a partial match with related dialogue contexts. For example, the proper dialogue contexts for a beg speech act to be selected are the emotion context: negative and the institutional relation context: subordinate-to-superio. That is to say for a virtual agent to *beg* someone, he should feel a negative emotion and he is subordinated to the person whom he is begging. However, an agent might still *beg* someone due to a private reason even if they did not experience the negative emotion.

In Figure 8.2, we illustrate the flow how an agent uses a speech act to change participators' mind states. First, agents choose a speech act according to his/her mental states. When the speech act is expressed, every participator can receive the speech act together with physical context information from the environment. In the ending phase, each participator's mental states could be affected as the effects of the speech act. Then the next speaker starts to choose another speech act in the next phase of dialogical interaction.

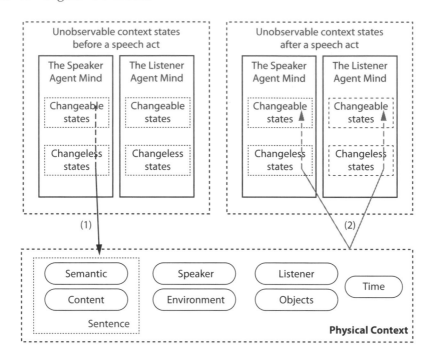

FIGURE 8.2 A speech act affects both speaker's and audience's mind contexts.

8.4 MULTIPLE DIALOGUE CONTEXTS

In the work of speech act classification [2], approximately 4,800 speech acts and 600 categories were divided into four major layers: expression, appeal, interaction, and discourse. Each speech act can be defined by the changing or triggering of specific contexts. It is impossible to list all applicable context conditions for a speech act since it falls into the frame problems [18], where all contingent conditions involved could not be specified. The ramification effects of a speech act on the mind states of other agents cannot be easily and clearly framed, due to the unobservable property of their mental states.

In designing the dialogue context model, we only considered the major conditions and effects that an independent speech act can achieve in a dialogue. Some principles and commonly encountered contexts, while considered arbitrary, can still be adopted in the dialogue context modeling to render the context awareness feasible. We often encounter the matter of the degree of effect in describing the relationship between the dialogue context and a speech act; to distinguish whether an agent is more likely to choose a particular speech act over another under a certain context, or whether a particular speech act is more likely than others to affect a given context. Our solution is to divide the degree of effect into five levels. Each level maps a probability in the Bayesian network, as follows: Level 1: 0.15, Level 2: 0.35,

Level 3: 0.5 , Level 4 : 0.65, and Level 5: 0.85. We then subjectively annotate the information into each speech act. Although the subjective annotation can cause inaccurate predictions in the beginning, we can later adjust the degree of effect dynamically at a separate learning stage based on the dialogue records, when the inaccurate predictions are found. For example, when we find that all agents tend to have low estimation toward the emotion anger, we could raise the levels of the effects of all speech acts that have affected the emotion anger.

8.4.1 Physical Context

Physical context describes some signals that can be detected by any kinds of sensors, and some signals that are provided by virtual environments and multi-agent systems. In agent dialogue, physical context contains several attributes of dialogues, such as location, time, who is the speaker and who is the listener, volume of speeches, objects and observable activities in the virtual environment, and sentences in dialogues.

8.4.2 Emotion Model

The OCC emotion model [16] proposed an emotion model for 22 types of emotions, according to their triggering conditions, in terms of an agent's appraisal of objects, agents, and events, in respect to their utility. However, some difficulties are encountered when using the OCC model to logically describe the emotion context for speech acts. The major difficulty is we must discriminate first the positive and negative emotion contexts for an agent.

In speech act classification [2], 155 speech acts have been identified under expression layer, which are used to express an agent's self emotion. In dialogue context, expressing self emotions can play a crucial role in agent communication, which can help an agent make other agents understand his/her reactions to previous conversations or to release emotional pressure [7], [14]. We manually divided those Expression Layer speech acts [2] into 2 groups, one group contains the speech acts with positive emotion context, and the other with negative emotion context.

8.4.3 Social Relation

Social relations refer to friendship/enemy relations. We focus on those relations in terms of two aspects that must be specified in the speech act: the relations that can be suggested or implied when a given speech act is adopted, and the relations that can be affected by the speech act.The friendship/enemy relations can affect the speech act at different degrees of influence toward the emotions of the listener agent. For example, if a friend pointing out your fault would probably provoke less anger from you than an enemy. It means if we monitor the long term emotion contexts changes in the dialogue, we may infer agents' friendship/enemy relations context.

8.4.4   Social Role

Some speech acts can only be used by agents in certain social roles. Conversely, when a speech act is issued by an agent, other agents might believe that the speaker agent actually plays the social role implied by the speech act. Institutional roles describe the authority and the obligation from official roles in an organization. In the speech act classification, some speech acts can only be used by agents in certain institutional roles. Conversely, when a speech act is issued by an agent, other agents might believe that the speaker agent actually plays the institutional role implied by the speech act. For an example, order is the speech act that can only be applied by an agent with a superior role to an agent with a subordinate role.

8.4.5   Personality

Personality specifies the tendencies of the reactions of an agent toward certain emotions and the violation of certain social norms. Since Personality is relatively unchangeable, without affecting the performance evaluation of other context models in this paper we could assume it is fixed at some prior constant for all dialogue agents.

8.4.6   Preference

In the preference context, we record the frequencies of speech acts used by a particular agent toward other agents under certain dialogue situations so that the tendency of choosing a particular speech act for the agent can be analyzed for future prediction. Similar to Personality context, we could assume preferences are also fixed at some prior constants for all dialogue agents.

8.5   SPEECH ACT MODEL WITH MULTIPLE CONTEXTS

Triggering contexts and effects describe dialogical context states in speech act model. Since dialogical context states are hidden in agent mind and can't be observed by others, we adopt a probability model to indicate the possible state of each dialogical context. For example, anger(0.7) means *the anger emotion has probability 0.7 to be true*. In this way, we can define triggering context and effects of each speech act with probabilistic dialogical context states. For instance, trigger(threaten): anger(0.7 +) means *if the anger emotion has probability 0.7 or more to be true, then agents would use speech act: threaten in dialogue*.

In Table 8.1, we illustrate an example of the probability models of a speech act: threaten. In this example, agent A is the speaker, and the only one audience is agent B. This speech act has 3 triggering dialogical contexts, such as the anger emotion of agent A, the hope emotion of agent A (hope for agent B's fear, obedience or something else), and the negative social relation: be-enemy-with between agent A and agent B. The effects of threaten will affect the mind contexts of both agents A and B, such as increasing the negative social relation

TABLE 8.1   PROBABILITY MODELS OF SPEECH ACT:
A THREATENS B

Triggers	(trigger, threaten, 0.85, anger) (trigger, threaten, 0.65, hope) (trigger, threaten, 0.65, be-enemy-with)
Effects	(effect,theaten, 0.85, anger, agent[B]) (effect, threaten, 0.65, fear, agent[B])

between agent A and agent B, causing agent B to feel fear, or causing agent B to feel angry. With the probability model, agent A could threaten agent B because agent A felt angry, or he/she looked for agent B's obedience, or both of them. In the same way, those effects are not inevitable, they are only possible.

TABLE 8.2   MAJOR SPEECH ACTS AND EMOTION EXPRESSIONS

Speech act	Accept Beg Grumble Order Reject Inform	Accuse Censure Interpellate Praise Reply Open dialogue	Agree-with Close dialogue Interrogate Propose Request	Ask Contorvert Not-intimidated Recount Threaten
Emotion	Cry Sad	Laugh	Happy	Rage

We implemented other 27 speech acts in a similar manner, to describe their triggering and effect conditions in the probability speech act model. Table 8.2 lists all 27 speech acts which occurred in our experiments. Those 27 speech acts include 22 major speech acts, and another type of speech act called emotion expression. This kind of speech acts has only triggering contexts and no effects. Because this type of speech acts, such as expressing sad, happy, rage, laugh, cry and so on, purely expresses speaker's emotion states. In this way, there are no reasonable effects can be defined in emotion expression.

8.6   CONTEXT AWARENESS IN DIALOGUE

Dialogical context in an agent mind is hidden and can't be directly observed by other agents. However, it can be inferred by observable actions and speech acts which are performed by agents. For example, we have no idea about a person's feeling if he/she didn't do anything. But once we observe the person is drinking water, we can infer that the person must feel thirsty. In the same way, we can infer a person's feeling, thought, intention from the person's speeches in conversation. However, to infer an agent's mind is not easy because of uncertainty. In our probabilistic speech act model, each speech act has multiple triggers and effects. For a speech act to be conducted, it is not

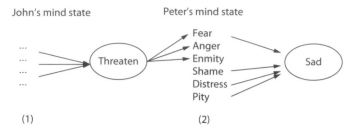

FIGURE 8.3 *John threatens Peter and Peter feels sad*; the emotion *fear* is one of the effects of threaten, and also one of the triggering contexts of "sad".

necessarily that all of triggers of the speech act are guaranteed to be satisfied and not all effects of the speech act will become true either. So it's risky to infer agent mind based on only very few speech acts. Agents might require history information as background or need a longer speech acts sequence to gather enough evidence to ensure better inference accuracy on various dialogue contexts.

In Figure 8.3, a scenario as "John threatens Peter and Peter feels sad" is shown. Figure 8.3 illustrates how the system can infer the emotion context "fear" by two consecutive speech acts. With the effects of the speech act: "threaten", we can understand Peter may raise emotions: "fear" or "anger" at the stage (2), and also may deem John as an enemy. By now, we can't be sure which effects will be achieved. However, we know those three effects are more likely to be true than the rest of dialogue contexts. Using our probabilistic speech act model and conduct Bayesian inference, all probabilities of these three effects are increased. On the other hand, supposed Peter expressed "sad" at the second stage. Sad is a speech act for emotion expression, and has no specific effect. However, "sad" can be triggered by several dialogical contexts, such as emotions fear, shame, distress and pity as well. If we don't retrospect the previous speech act, we can't make sure which triggering conditions of the dialogue context "sad" are true. Similar to the previous stage, all the probabilities of triggers of the dialogue context "sad" are increased by Bayesian inference. Now we notice that the dialogue context states at stage (2), the emotion "fear" is actually the only dialogue context state that is increased twice by both speech acts. Therefore, the system can infer that Peter most likely feels fear at the stage (2).

8.7 COMPUTATIONAL MODEL OF CONTEXT AWARENESS REASONING USING DBN

In DBN modeling and implementation [15], we must generally specify the domain sensor model and transition model in terms of conditional probabil-

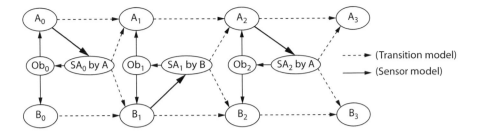

FIGURE 8.4 Two DBN models for both the speaker's and the listener's dialogue contexts. The speech act and other observation information are the sensor model in DBN.

ities P $(E_t|X_t)$ and P $(X_{t+1}|X_t)$, respectively, where E_t represents the evidence collected from sensors at time t, and X_{t+1} and X_t represent the domain states at time t+1 and t. With an assigned initial state P (X_0), we can obtain the states of X at any t as in equation 8.1.

$$P(X_{0:t}, E_{1:t}) = P(X_0) \prod_{i=1}^{t} P(X_i|X_{i-1})P(E_i|X_i) \tag{8.1}$$

Since each speech act is modeled by change of dialogue contexts, we could subjectively attach the probability of the possible change of context preconditions and post conditions in the speech act as a transition model. The sensor model gathers evidence from the physical context, which includes the content semantics of the dialogue sentence and other information cues via observations, such as tone of speech, facial expression, gesture, and object. In Figure 8.7, we showed two DBN models for two participants, agent A and agent B, during conversation. A_t and B_t are unobservable dialogue contexts for their mental states in dialogue step t. The observable data (Ob_t), which include the speech act (SA_t), are the sensor model that reflects states of observable dialogue contexts evidence E_t. The speech act could also affectthe dialogue context. The effects of speech act SA_{t-1} could cause the dialogue context to change from X_{t-1} to X_t with probability P $(X_t|X_{t-1}, SA_{t-1})$ as a transition model. Using the speech act models, we can compute the probabilities on the states of X at dialogue step t according to equation 8.2 by combining the sensor model and transition model as in a DBN model.

$$P(X_{0:t}, E_{1:t}) = P(X_0) \prod_{i=1}^{t} P(X_i|X_{i-1}, SA_{i-1})P(OB_i|X_i) \tag{8.2}$$

Although DBN is such a power computational model to acquire probability predictions for context awareness, it creates another problem that we must solve is time-consuming. In our experiment, we implemented 47 dialogue contexts as DBN model, and it costs 60 to 300 seconds to accomplish a cycle of computation for one speech act input. This disadvantage is critical in real-time interactive virtual dramas and games. To overcome this problem, we combined two methods with our DBN model: Noisy-OR and Isolated contexts.

8.7.1 Noisy-OR Model

To combine the related conditional probability, we use a noisy-or model, under the assumption that all the contexts with conditional probabilities are independent. The idea behind Noisy-OR [8], [18] function is that a speech act SA with n trigger contexts C_i, each with a probability value p_i, where p_i is the probability that SA = true on C_i = true while C_j = false $\forall$ j $\neq$ i as in equation 8.3.

$$p(SA = true | C_1, ..., C_n) = 1 - \prod_{i:C_i=true} (1 - p_i) \tag{8.3}$$

The limitation of using the noisy-or model to calculate the CPT is that we can only design triggers for a speech act. We cannot describe the type of condition in which a context might reduce the possibility of invoking a speech act with the noisy-or model. To use the noisy-or model, the assumption that all dialogue contexts are independent can be substantial. However, using the noise-or model, we could simplify the computation complexity of calculating the conditional probabilities table (CPT), by reducing all 2n combinations of true-false possible conditions to only n-item computation. The Noisy-Or method can reduce one cycle computation time into several seconds from hundreds seconds.

8.7.2 Isolated Contexts

We can tolerate several seconds of delayed response in real-time games, but it's not good enough obviously. According to statistics, each speech act input will only change about 3 to 7 condition probabilities in dialogue contexts. In the previous model, we have total 47 dialogue contexts and all of them must cost at least one calculation in one DBN cycle. In other words, most of computation resources are taken by irrelevant dialogue contexts (according to the input speech act). So we design an isolated context for each speech act to avoid unnecessary calculation. In order to use this isolated context correctly, two guiding rules must be followed before we skip the calculation for a dialogue context: 1. The probability of current dialogue context is under 1% to the dialogue context with highest probability. 2. The skipped calculation will not raise the probability of the dialogue context. By using isolated contexts method, we save 61% 73% time cost in the whole dialogue script computation.

8.8 EXPERIMENT

To demonstrate our approach works in games, we use multi-agent system and agent technology to play some virtual dramas. Our goal is to prove agents can have ability to aware context states in dialogue with our speech acts model and DBN computational method.

JACK multi-agent platform [4] was chosen as our multi-agent system because it has the integrated agent communication function, and we can design agent behaviors with simple scripts. We also obtained several dialogical scripts from famous movies, and made intelligent agents as non-player characters (NPCs) to play the scenario. Each agent has their own dialogue script which includes dialogue sentences, transferred speech acts and targets to speak. We also pre-defined personality, preference and other mind states to each individual NPC by referring the background setting in the original movie scenarios. Then we made NPCs to play these scenarios automatically, and to infer other participants' mind states with our speech act model and DBN. Meanwhile, we annotated dialogical context states to each sentence in the scenarios by several people. Afterward, we compare human-annotated data and NPC-inferred results, and show the analysis in sections 8.8.1 and 8.8.2 respectively.

We obtained several dialogue sequences from the scripts of famous films "Doubt" [19], "The King's Speech" [13], "Pride and Prejudice" [6] and "Meet Joe Black" [3]. There are two reasons for choosing these scripts. First, obvious emotion changes are present in these scripts, and it's beneficial for us to verify NPC's inference. Second, those emotion changes are by speech acts, not by behaviors during the dialogue, so physical action effects can be ignored in the scenarios.

8.8.1   A Speech Act Model with Multiple Contexts

Table 8.3 lists 6 major types of dialogical context in the experiment. There are total 47 dialogical context states can be inferred and aware by agents in dialogue. Without loose of generality, we treated some of them as background information and initial conditions, such as social roles, dialogue games and social relations. The reason is that our scenarios were extracted from the middle part of movies, so characters have already encountered and interacted each other for some time.

Although our dialogical emotion contexts is based on OCC emotion model, due to the limited effort and computation, as well as the test bed domain for obtaining the full performance, we only focus on 8 emotion sets: (shame, pride), (remorse, gratification), (fear-confirmed, satisfaction), (reproach, admiration), (disappointment, relief), (anger, gratitude), (fear, hope) and (hate, love). We choose the negative emotions to stand for those emotion sets. For example, shame(0.7) means the emotion shame has a probability 0.7 to be true, and also means the emotion pride has a complementary probabil-

TABLE 8.3   6 DIALOGICAL CONTEXTS

Emotion contexts	Social relations	Social roles
Shame/Pride	Be-enemy-with	Father
Disappointment/Relief	/Be-friend-with	Sister
Anger/Gratitude	Be-authority-to	God
Hate/Love	/Be-Subordinate-to	Mortal
Remorse/Gratification		Lady
Fear-confirmed/Satisfaction		Gentlemen
Reproach/Admiration		King
Fear/Hope		Civilian
Personality	Preference	Dialogue games
Solemn	Information	Propose/Cancel
Conservative	Peace	Inform/Cancel
Gentle	Keep Secret	Ask/Cancel
Self-Confident	Authority	Request/Cancel
Impassive	Be accepted	
Self-doubt		

ity 0.3 to be true. The reason for choosing negative emotions is the scenario we chose primarily presents debating arguments. So we tend to model negative emotions to reveal the experimental results more clearly. In addition to emotion context, we also model several social relation contexts, such as be-enemy-with/be-friend-with to model the friendship relations, and a set of paired relations be-authority-to/be-subordinate-to to model social hierarchical relations. The latter is one of the triggering contexts of the speech act: order. Social roles father, sister, god are implemented according to the movie character domain.

8.8.2   Infer Social Context from Dialogue

In the experiment, we attempt to assess if the speech act model could correctly infer the emotions and relations of the two characters. In the scene from the movie "Doubt", Sister Aloysius interrogates Father Flynn about the child abuse from the beginning. With solemn and conservative personality, Sister Aloysius tends to hide negative expressions in the conversation. Although lacking of direct evidence of negative emotion from pure emotional speech acts, the NPC Father Flynn can still infer her emotion context states. It is because Sister Aloysius frequently uses several aggressive speech acts, such as interrogate, reject, accuse, and controvert. Those speech acts are clues of Sister Aloysius' enmity toward Father Flynn. In Figure 8.5, we showed that Sister Aloysius had enmity to Father Flynn from the beginning to the end, and she also felt anger during the conversation. Even if she pretended she was calm and friendly sometimes (at sentence #1, #5, #6, #8, #19, #20 and #26). After DBN processing, we made agents have ability to realize the truth.

The result in Figure 8.5 shows Sister Aloysius' emotion: anger and social

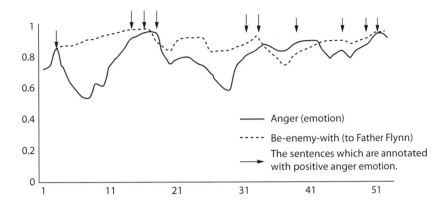

FIGURE 8.5 Inferred Sister Aloysius' emotion: anger and social relation be-enemy-with (to Father Flynn). The solid line is the baseline which is annotated by several people.

relation context: be-enemy-with (Father Flynn). Although most part of those two dialogical contexts have similar trend, they still have obvious difference in sentences #3-6 and #26-29. The reason is that Sister Aloysius didn't use aggressive speech acts, so the inferred emotion: anger is at average level. However, at the same time, she tried to ask questions to Father Flynn, but he didn't give any answer. This kind of uncooperative behaviors in dialogue raised Sister Aloysius' enmity. It's why she didn't look like very angry from facial expression, but she actually possessed high enmity to Father Flynn from dialogue.

To verify the inferred result more broadly, we annotated several dialogical contexts from original movies as baselines by several people. We ask annotators annotated those dialogue sentences that can be judged by facial expression, tone, gesture and other clues from the original movie which are indicated as down arrows in Figure 8.5. In this way we obtain the baselines that avoid the situations of having to infer character's mental states. In Figure 8.5, if the actor or the actress performed an angry facial expression, or spoke with a raging tone, then we annotate "anger(1)" to the sentence. Similarly, we annotate "anger(0)" when the actor performs gratitude to someone. Otherwise, annotate "anger(0.5)" to ambiguous or emotionless expressions.

The DBN inference results are indicated as dashed (social relation of Sister Aloysius to Father Flynn) or solid curves (emotion levels of Sister Aloysius) in Figure 8.5. Compare the inferred results with the baseline, we can observe that the inferred results and the baseline are pretty matched. The rising trend of emotion anger is inferred correctly almost at every sentence which is also annotated the same emotion expression in baseline with a down arrow.

In the same way, our DBN model also inferred emotion anger in other 3 scenarios, and we compare the results with the baselines as shown in Figure 8.6. There are only 3 unmatched results in our experiment in inferring emotion

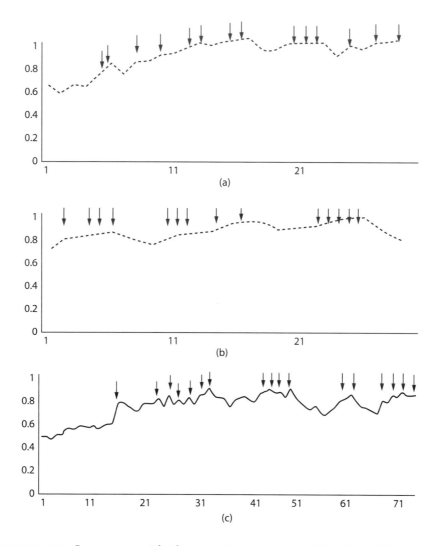

FIGURE 8.6 Compare with the emotion: anger (with dotted line) and the baseline (with solid down arrows) in three scenarios. (a) Bill Parrish in the movie "Meet Joe Black", (b) Lizzy in the movie "Pride and Prejudice" and (c) King George VI in the movie "The King's Speech".

TABLE 8.4   MATCHING SCORES IN OTHER EMOTION CONTEXTS

Emotion	Score	Baseline	Inferred (rise)
Shame/Pride	73.33%	33	45
Remorse/Gratification	27.27%	6	22
Disappointment/Relief	76.60%	36	47
Fear-confirmed/Satisfaction	40.90%	9	22
Anger/Gratitude	92.86%	39	42
Fear/Hope	86.79%	46	53
Hate/Love	20.51%	8	39
Reproach/Admiration	59.65%	34	57

TABLE 8.5   FOUR BASIC SPEECH ACT TYPES AND THEIR ELABORATED SPEECH ACTS IN DOUBT'S SCENARIO

Reguest	Recount	Ask	Reply
Request	Recount	Ask	Reply
Order	Censure	Interrogate	Reject
Propose	Accuse	Interpellate	Accept
		Threaten	Controvert
			Threaten
			Not-intimidated

context: anger. The matching score is 92.86% which is the best score within experiments. Table 8.4 shows each matching score in other dialogical emotion contexts. Although the best score is 92.86% (anger), we still have 4 emotions scores beyond under 60%. The reason is that there is no semantic content of sentence so far in our speech act model. Agents can't judge good/bad or right/wrong on the objects, events and other agents' behaviors, from the content of sentences. The other shortcoming in our experiments is, even though we can infer any dialogical context in our model, due to lack of baseline model, we can only score the emotion context at this stage.

8.8.3   Noise in Dialogue Context Awareness

We mentioned that agents should have the ability to communicate without Grice's maxims. That is, agents might lie or hide their true intentions in dialogues. So a listener agent could thus get noisy speech acts that affect the accuracy of context awareness. To evaluate such noisy effect on the context awareness from dialogue using DBN inference, we designed an experiment that allowed agents to replace their speech acts with basic type of speech acts. In table 8.5, we define four kinds of noisy speech act classification and each can replace three to six true speech acts according to its context. Speech acts in the same category have the same enaction type, so they transfer the same information in the dialogue without leaking out any other dialogue contexts, such as emotion and preference, to the listener.

TABLE 8.6 ORIGINAL INPUT SEQUENCE AND ONE WITH NOISY SPEECH ACTS

input with correct speech acts	input with noisy speech acts
(21a)(12b)(21c)(21b)(13e)(31f)	(21a)(12b)(21c)(21d)(13e)(31f)
(21g)(12e)(21f)(12h)(21g)	(21e)(12e)(21f)(12f)(21e)

Table 8.6 is the input dialogue sequences in terms of the basic types of speech acts. The annotations of semantic speech act sequence and pragmatic speech act sequence, corresponding to the 53 dialogue sentences in the dialogue script, are annotated as A and B, respectively. Each dialogue character is labeled with a number: 1 for Father Flynn, 2 for Sister Aloysius, and 3 for Sister James. Speech acts are labeled with a symbol from a to u in which: a: Announce, b: Controvert, c: Dissatisfied, d: Recount, e: Ask, f: Reply, g: Interrogate, h: Reject, i: Censure, j: Rebut, k: Say-goodbye, l: Accuse, m: Agree-with, n: Request, o: Propose, p: Be-glad, q: Threaten; r: Not-intimidated, s: Pride, t: Praise, and u: Grumble. Ignoring sentence content semantics for this experiment, each dialogue sentence can be abbreviated as speaker + audience + speech act. For example, Sister Aloysius makes an announcement speech act (a) to Father Flynn, which will be annotated as (21a), in the semantic speech act sequence. We replaced 13 speech acts to the basic speech acts at sentences #7, #10, #11,#12, #13, #23, #27, #29, #30, #36, #38, #39 and #40.

In Figure 8.7, we showed the anger, fear, and disappointment emotion states of Father Flynn during original conversation. It is apparent that Father Flynn has a high anger and fear emotion level while talking with Sister Aloysius. At the begging of the conversation, Sister Aloysius said, "I'm not satisfied that that is true" (#3 means the third sentence in the scenario) after Father Flynn said, "His well-being is not at issue"(#2). In this scenario, we observe that the "fear" emotion is rapidly increasing. The pragmatic speech act "controvert" tends to arouse the "fear" emotion in speakers in light of the disapproval of others. As Aloysius "dissatisfies" and "controverts" Flynn's argument, the "fear-confirmed" emotion level of Flynn will increase, due to the evidence of pragmatic speech acts sequence. This evidence is modeled in "Discourse Act" context, to represent the sequence of speech acts that could also have effects on other context states.

In Figure 8.8, we showed the anger, fear, and disappointment emotion states of Father Flynn during conversation with the noisy input sequence. We can notice Father Flynn's negative emotions are not as high as the previous experiment. That is because most aggressive speech acts are replaced with basic types. However, we still can detect his short term emotion from facultative aggressive speech acts (in the sentence #2, #3, #16, #17, #24, #32, #35, #50) which is not replaced. In the end of the dialogue, because Sister Aloysius did not threaten Father Flynn anymore in the noisy input, so Father Flynn's fear emotion is decreased to very low level. (Maybe this is actually what Father Flynn wants to represent in the dialogue: hiding his fear). But Sister Aloy-

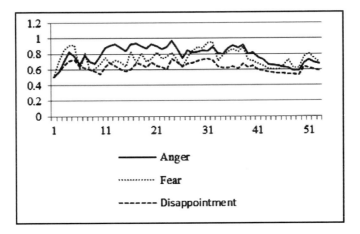

FIGURE 8.7 Father Flynn's anger, fear and disappointment state in Emotion context with original speech act input sequence.

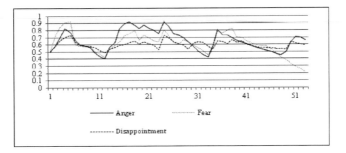

FIGURE 8.8 Estimated Father Flynn's anger, fear and disappointment state in Emotion context with noisy input sequence.

TABLE 8.7   SEMANTIC AND PRAGMATIC SPEECH-ACT SEQUENCES IN
DOUBT SCENARIO

Semantic speech acts input sequence	Pragmatic speech acts input sequence
(21a)(12b)(21c)(21d)(13e)(31f)(21e)	(21a)(12b)(21c)(21b)(13e)(31f)(21g)
(12e)(21f)(12f)(21e)(12f)(21e)(32e)	(12e)(21f)(12h)(21g)(12h)(21g)(32e)
(23f)(12i)(21j)(12e)(21f)(21e)(12e)	(23f)(12i)(21j)(12e)(21f)(21g)(12e)
(21f)(21e)(12h)(12k)(13k)(21d)(12e)	(21f)(21g)(12h)(12k)(13k)(21l)(12e)
(31f)(21e)(12n)(21h)(31n)(12n)(21h)	(31m)(21g)(12n)(21h)(31n)(12n)(21h)
(12f)(31p)(21e)(12f)(12f)(31t)(13s)	(12b)(31p)(21q)(12r)(12q)(31t)(13s)
(31t)(32e)(23f)(12m)(12e)(21f)(12d)	(31t)(32e)(23f)(12m)(12e)(21f)(12d)
(21u)(12l)(12k)(12k)	(21u)(12l)(12k)(12k)

sius still Grumbled in the sentence #50, so Father Flynn felt anger due to the
speech act "Grumbled", and accused Sister Aloysius in the sentence #51.

8.8.4   Pragmatic Prediction with Dialogue Contexts

As discussed above, if all agents conduct dialogues by obeying Grice's maxims, we could easily convert the dialogue content semantics directly into
pragmatics (infer speech acts directly from their dialogue content semantics)
[5]. However, in emergent narratives, we do not have assumption of Grice's
maxims for agent communication, and therefore, could not directly obtain
the pragmatics or speech acts from the dialogue content semantics. Since
the agent's mental states are essentially unobservable, we cannot confirm
whether the pragmatics in a dialogue sequence of inferred speaker agents are
indeed the true intentions (or speech acts) of the speakers. Therefore, we distinguish between two different dialogue sequences: (1) a coherent semantic
sequence, in which all the dialogue semantics of speech acts in the dialogue
sequence can be treated as pragmatics and no conflict in dialogue contexts of
those speech acts can be found in the sequence; and (2) an incoherent semantic
sequence in which the dialogue semantics of speech acts are treated directly
as pragmatics, and if so, the dialogue contexts of certain speech acts might be
in conflict with their true pragmatics. Dialogues in the emergent narratives
are generally incoherent semantic sequences, and therefore, we must find an
explanation of the most likely pragmatic speech act sequence for a given dialogue sequence. To achieve this, we should make agent aware the dialogue
contexts in the agent conversation. In this experiment, we intend to show that
a dialogue agent can predict the correct pragmatic speech act to some extent
from its semantic speech act of a dialogue sentence in the agent dialogue
conversation given the dialogue contexts of the speech act model. In Table
8.7, we demonstrated that 21 speech acts occurred in the conversation. Part
(a) is semantic speech act input sequence, and part (b) is pragmatic speech
act input sequence.

We assume the correct pragmatic speech acts in the first half of the speech
acts in the script are given as known, and then each dialogue sentence is

TABLE 8.8   THE ACCURACY OF PRAGMATIC SPEECH ACT PREDICTION
WITH/WITHOUT PRELOADED CONTEXT

Correct data	lemgnfnnhbpqrqtstefmefdulkk	Accuracy
Results with context info	lemgnfnnhbpefftstefmefdudkk	15/19
Results w/o context info	demenfnnffpefftstefmefdudkk	11/19

input one by one as continuing the second half of the dialogue. We model
two different agents as a test. The first agent will be given the preloaded
context information in the first half part of dialogue log. By the pre-loaded
context information, we mean all the pragmatic speech acts in the dialogue
sequence have been conducted so far. We observe its prediction ability on the
pragmatics based on every dialogue sentence at the second half. The second
agent will not be given any preloaded context information, so he/she is the
third party agent and join in the conversation in the middle. Of course, we
expect the second agent to have a lower accuracy for prediction than the
first one as a contrast. In the Table 8.8, the first row is the pragmatic speech
act sequence for the second half part in the scenario. The second row is the
predicted results of the first agent with pre-loaded context information. The
third row is the predicted result from the second agent without pre-loaded
context information. The third column in row 2 and 3, we calculated the
accuracy ratios of precision of the two agents respectively. All the mismatches
are indicated in grey shade.

We only calculate the accuracy with the predicted result for the speech
act classification. There are 16 classified speech acts, and 5 speech acts are
not classified. It means that if a semantic speech act is not classified, the
pragmatic speech act will be equivalent to the semantic speech act. In this
experiment, only 19 of 26 sentences have classified semantic speech acts. The
first result in this experiment with preloaded context knowledge has four
error predictions with accuracy rate 15/19. The reason of error is due to, in
the end of the conversation, emotion intensity is at normal level, so agent
can't easily distinguish the pragmatic speech acts using Emotion context.
The second result shows a worse performance of an agent without context
information. However, it still has an accuracy rate of 11/19. It's because the
dialogue context used to predict the speech act pair "ask-reply" actually make
effects. In this experiment, we show that the accuracy with preloaded context
knowledge (namely, the accumulated context information during dialogue)
helps in predicting the pragmatic speech act from a semantic one.

8.8.5   The Most Likely Pragmatic Speech Acts Sequence

In this experiment, we assume only semantics of dialogue sentences are given
as known, we attempt to assess if the speech act model could find out the
most likely explanation of the dialogue context. Since most dialogue sentences
are "ask/reply" speech acts, but sometimes emotions of dialogue agents can

TABLE 8.9   THE PROBABILITY AND ACCURACY OF THE
MOST LIKELY PRAGMATIC SPEECH ACT SEQUENCE

Correct data	abcbefgefhghgefijefgefghkk lemgnfnnhbpqrqtstefmefdulkk	Probability Accuracy
Output Result	adcdefeeffghgefijefeefehkk lgmgnhnnhbpefftstefmefdudkk	6.792e-11 29/40

become incompatible with the contexts, we wish to know to what extent the model could find an explanation of pragmatic context (as "interrogation" in this case) for each dialogue speech act (as "ask" in this case) sentence. With the same reason mentioned in section 8.8.4, we calculate the accuracy based on the classified speech acts.

Using DBN, we calculate a most likely pragmatic speech act sequence from its corresponding semantic sequence whose overall probability is 6.792e-11 with 29 correct pragmatic speech act matches out of 40 semantic speech acts as shown in Table 8.9. We reason that the error could be due to the "peaceful" conversation at the beginning of the scenario that provides little emotional context. So the prediction of the pragmatic speech act interrogate from semantic speech act ask is incorrect at the beginning for about the first one third of conversation.

8.9   CONCLUSION

We developed an integrated probabilistic model to equip non-player characters with context awareness on multiple dialogue contexts, and tested it against an annotated baseline extracted from the scripts of the films "Doubt","Meet Joe Black", "Pride and Prejudice" and "The king's speech". This approach combines dynamic Bayesian network and speech acts, as well as background models on emotions and social relations. Even initialized with only partial background knowledge, our method is able to detect hidden states of emotions and social relations to some extent by only using speech acts. This implies even without sufficient background knowledge, we could still show the context awareness with proper speech act model in dialogue. Although initial probabilities are subjective and context knowledge is incomplete, we have demonstrated the feasibility of DBN dialogue model and speech act model to achieve context awareness from dialogue. We believe better performance can be tuned by incorporating more complete knowledge from different background theories and learning mechanisms, to refine parameters when prediction errors occur in the DBN model and by enhancement of efficiency of computational methods. Of course, more advanced natural language processing and understanding techniques to infer basic types of semantic speech acts from language sentences are also needed to be developed.

We offer two main contributions. First, our context awareness model shows the feasibility to enable agents or robots to reason unobservable social context states during dialogue interactions with other agents. Second, we develop models of context awareness devised on speech acts, which allows integration with a probabilistic planner that could generate dialogue plans for robots.

8.10   Bibliography

[1] Matthias Baldauf, Schahram Dustdar, and Florian Rosenberg. A survey on context-aware systems. *Int. J. Ad Hoc and Ubiquitous Computing*, 2(4):263–276, 2007.

[2] T. Ballmer and W. Brennenstuhl. *Speech Act Classification: a Study of the Lexical Analysis of English Speech Activity Verbs*. Springer-Verlag, Berlin; New York, 1981.

[3] M. Brest, R. Osborn, and J. Reno. Meet joe black. http://www.imdb.com/title/tt0119643/, 1998.

[4] Paolo Busetta, Ralph Ronnquist, Andrew Hodgson, and Andrew Lucas. JACK Intelligent Agents - Components for Intelligent Agents in Java. *AgentLink News*, (2), 1999.

[5] Alex Yu-Hung Chien and Von-Wun Soo. Inferring pragmatics from dialogue contexts in simulated virtual agent games. In *In Proceedings of AEGS 2011 Workshop, LNAI7471*, pages 123–138. Springer-Verlag Berlin Heidelberg, 2011.

[6] C. Coke, J. Austen, and F. Weldon. Pride and prejudice. http://www.imdb.com/title/tt0078672/, 1980.

[7] C. Conati. Probabilistic Assessment of UserâĂŹs Emotions in Educational Games. *Applied Artificial Intelligence*, 16:555–575, 2002.

[8] F. G. Cozman. Axiomatizing noisy-or. In *In Proceedings of European conference on artificial intelligence*, pages 979–980. IOS Press, Valencia, Spain, 2004.

[9] Anind K. Dey and Gregory D. Abowd. Towards a Better Understanding of Context and Context-Awareness. In *HUC '99 Proceedings of the 1st International Symposium on Handheld and Ubiquitous Computing*, pages 304–307. Springer-Verlag London, UK, 1999.

[10] T. Finin, R. Fritzson, D. McKay, and R. McEntire. Kqml as an agent communication language. In *Proceedings of the 3rd International Conference on Information and knowledge management*, page 456âĂŤ463. ACM. Gaithersburg, Maryland, 1994.

[11] Foundation for Intelligent Physical Agents (FIPA). Fipa communicative act library specification. *FIPA00037*, 2014.

[12] P. Grice. *Studies in the Way of Words*. Harvard University Press, 1989.

[13] T. Hopper and D. Seidler. The kingâĂŹs speech. `http://www.imdb.com/title/tt1504320/`, 2010.

[14] Z. Inanoglu and R. Caneel. Emotive alert: Hmm-based emotion detection in voicemail messages. In *In Proceedings of the International Conference on Intelligent User Interfaces*, pages 251–253. ACM. San Diego, California, USA, 1995.

[15] K. P. Murphy. *Dynamic Bayesian Networks: Representation, Inference and Learning*. Ph.D. Dissertation, UC Berkley, USA, 2002.

[16] A. Ortony, G. L. Clore, and A. Collins. *The Cognitive Structure of Emotions*. Cambridge University Press, 1988.

[17] Anand S. Rao and Michael P. Georgeff. Bdi agents: From theory to practice. In *In Proceedings of the first International Conference on Multi-Agent Systems (ICMAS-95)*, pages 312–319, 1995.

[18] S. Russell and P. Norvig. *Artificial Intelligence: A Modern Approach*. Prentice Hall, 3rd edition, 2009.

[19] J. P. Shanley. Doubt, 2008. `http://www.imdb.com/title/tt0918927/`.

VI

Psychological Aspect of Cognitive Robotics

Robotic Action Control: On the Crossroads of Cognitive Psychology and Cognitive Robotics

Roy de Kleijn

Leiden Institute for Brain and Cognition,
Leiden University, The Netherlands.

George Kachergis

Leiden Institute for Brain and Cognition,
Leiden University, The Netherlands.

Bernhard Hommel

Leiden Institute for Brain and Cognition,
Leiden University, The Netherlands.

CONTENTS

T HE field of robotics is shifting from building industrial robots that can perform repetitive tasks accurately and predictably in constrained settings, to more autonomous robots that should be able to perform a wider range of tasks, including everyday household activities. To build systems that can handle the uncertainty of the real world, it is important for roboticists to look at how humans are able to perform in such a wide range of situations and contexts–a domain that is traditionally the purview of cognitive psychology. Cognitive scientists have been rather successful in bringing computational systems closer to human performance. Examples include image and speech recognition and general knowledge representation using parallel distributed processing (e.g., modern deep learning models).

Similarly, cognitive psychologists can use robotics to complement their research. Robotic implementations of cognitive systems can act as a "computational proving ground", allowing accurate and repeatable real-world testing of model predictions. All too often, theoretical predictions–and even carefully-conducted model simulations–do not scale up or even correspond well to the complexity of the real world. Psychology should always seek to push theory out of the nest of the laboratory and see if it can take flight. Finally, cognitive psychologists have an opportunity to conduct experiments that will both inform roboticists as they seek to make more capable cognitive robots, and increase our knowledge of how humans perform adaptively in a complex, dynamic world. In this chapter, we will give a broad but brief overview of the fields of cognitive psychology and robotics, with an eye to how they have come together to inform us about how (artificial and natural) actions are controlled.

9.1 EARLY HISTORY OF THE FIELDS

9.1.1 History of Cognitive Psychology

Before cognitive psychology and robotics blended into the approach now known as cognitive robotics, both fields already had a rich history. Cognitive psychology as we now know it has had a rocky past (as have most psychological disciplines, for that matter). Breaking away from philosophy, after briefly attempting to use introspection to observe the workings of the mind, the field of psychology found it more reliable to rely on empirical evidence.

Although making rapid strides using this empirical evidence, for example in the form of Donders' now classic reaction time experiments which proposed stages of processing extending from perception to action, early cognitive psychology came to be dominated by a particular approach, *behaviorism*. This position, popularized by Watson [54] and pushed further by Skinner [46], held that the path for psychology to establish itself as a natural science on par with physics and chemistry would be to restrict itself to observable entities such as stimuli and responses. In this sense, behaviorists such as Skinner were strongly anti-representational, i.e., against the assumption of internal knowledge and states in the explanation of behavioral observations. On the other hand, the focus on observable data brought further rigor into the field, and many interesting effects were described and explained.

The behaviorist approach dominated the field of psychology during the first half of the 20th century. In the 1950s, seeming limitations of behaviorism fueled what some scholars would call the *neocognitive revolution*. Starting with Chomsky's scathing 1951 review of Skinner's book that tried to explain how infants learn language by simple association, many researchers were convinced that behaviorism could not explain fundamental cognitive processes such as learning (especially language) and memory. The foundations of the field of artificial intelligence were also nascent, and pursuing explanations of high-level, uniquely human aptitudes–e.g., analytical thought, reasoning, logic, strategic decision-making–grew in popularity.

9.1.2 The Computer Analogy

Another factor contributing to the neocognitive revolution was the emergence of a new way to describe human cognition as similar to electronic computer systems. The basic mechanism operating computers was (and still is, in a fundamental way) gathering input, processing it, and outputting the processed information, not unlike the basic cognitive model of stimulus detection, storage and transformation of stimuli, and response production.

Clearly, this processing of information requires some representational states which are unaccounted for (and dismissed as unnecessary) by behaviorists. This new way to look at human cognition as an information processing system not only excited psychologists as a way of understanding the brain, but the analogy also raised hopes for building intelligent machines. The idea was that if computer systems could use the same rules and mechanisms as the human brain, they could also *act* like humans. Perhaps the most well-known proponent of this optimistic vision was Turing [51], who suggested that it wouldn't be long before machine communication would be indistinguishable from human communication. Maybe the secret of cognition lies in the way the brain transforms and stores data, it was thought.

Alas, the optimists would be disappointed. It soon became clear that computers and humans have very different strengths and weaknesses. Computers can calculate pi to twenty significant digits within mere milliseconds. Humans can read terrible handwriting. Clearly, humans are not so comparable to basic input-output systems after all. It would take another 25 years for cognitive psychology and artificial intelligence to begin their romance once again, in the form of the *parallel distributed processing* (PDP) approach [40].

9.1.3 Early Cognitive Robots

With this idea of smart computer systems in mind, it seemed almost straightforward to add embodiment to build intelligent agents. The first cognitive robots were quite simple machines. The *Machina Speculatrix* [53] consisted of a mobile platform, two sensors, actuators and 'nerve cells'. Understandably, these robots were designed to mimic behavior of simple animals, and could move safely around a room and recharge themselves using relatively simple approach and avoidance rules.

Due to their simplicity, it was questionable exactly how *cognitive* these robots were–they are more related to cybernetics and control theory (e.g., [5])–but soon enough complexity made its way into cognitive robotics.

From the 1960s, robots would be able to represent knowledge and plan sequences of operations using algorithms such as *STRIPS* [17], that would now be considered essential knowledge for every AI student. The STRIPS planner, which represents goal states and preconditions and attempts to derive the action sequences that would achieve them before carrying them out, is quite slow to execute. Moreover, this type of planning suffers from its closed world assumption (i.e., that the environment and all relevant states are known–by programming–and will not change), and the massive complexity of the real world, leading to intractable computations. Yet the general approach taken by STRIPS–of modeling the environment, possible actions and state transformations, and goal states via predicate logic, and operating robots via a sense-plan-act loop–has dominated cognitive robotics for quite some time, and is still a strong thread today.

Various behavior-based robotics architectures and algorithms–taking some inspiration from biological organisms–have been developed in the past few decades. An early, influential example is Rodney Brooks' subsumption architecture [9], which eschews planning entirely–"planning is just a way of avoiding to figure out what to do next", using a defined library of basic behaviors arranged hierarchically to generate behavior based on incoming stimuli. Although fast and often generating surprisingly complex behavior from simple rules (see also [8]), the subsumption architecture and many other behavior-based robotics algorithms do not yet incorporate much from the lessons to be learned from psychological studies in humans.

9.2 ACTION CONTROL

9.2.1 Introduction

One of the other areas that shows considerable overlap between robots and humans is motor/action control. Two types of control systems govern motor action: *feedforward* and *feedback* control systems.

A feedforward motor control system sends a signal from the (human or robotic) motor planning component to the relevant motor component using predetermined parameters, executing said action. Information from the environment can be considered only before execution begins, which makes feedforward control suitable for predictable environments.

In contrast, a feedback motor control system incorporates information from itself or the environment (feedback) more or less continuously to modulate the control signal. In this way, the system can dynamically alter its behavior in response to a changing environment.

9.2.2 Feedforward and Feedback Control in Humans

For many years, psychology and related disciplines have approached action control from rather isolated perspectives. As the probably first systematic study on movement control by Woodworth [55] had provided strong evidence for the contribution of environmental information, many authors have tried to develop closed-loop models of action control that rely on a continuous feedback loop (e.g., [1]). At the same time, there was strong evidence from animal and lesion studies [31, 49] and from theoretical considerations [34] that various movements can be considered in the absence of sensorimotor feedback loops, which has motivated the development of feedforward models (e.g., [22]).

Schmidt [43] was one of the first who argued that human action control consists of both feedforward and feedback components. According to his reasoning, human agents prepare a movement schema that specifies the relevant attributes of the intended movement but leave open parameter slots that are specified by using online environmental information. Neuroscientific evidence has provided strong support for such a hybrid control model, suggesting that off-line action planning along a ventral cortical route is integrated with online sensorimotor specification along a dorsal route [19, 18].

In particular, feedforward mechanisms seem to determine the necessary action components off-line and pre-load at least some of them before initiating the action [22], and to selectively tune attention to stimuli and stimulus dimensions that are relevant for the task [24]. Feedback processes, in turn, provide excellent accuracy–often at the cost of speed [44]. These strengths and weaknesses have motivated hybrid models claiming that feedforward mechanisms provide the skeleton of action plans which leave open slots for parameters provided by feedback processes [43, 18, 24].

A particularly good example of this kind of interaction is provided by the observations of Goodale and colleagues [20]. In a clever experiment, participants were asked to rest their hand on a platform and point to a visual target presented at a random location on an imaginary line in their right visual field. The participants were not told that in half of the trials the target changed location during the first saccade. The authors found that participants would successfully point to the target on these trials without even being aware of the location change, and without additional delay. As feedforward programming is thought to take time, a fast and online feedback mechanism of which participants are unaware has to be responsible for this finding.

On a higher level, interaction between feedforward and feedback systems must exist for goal-directed action to be carried out. Higher level, goal-directed action planning, such as planning to make pancakes would be impossible to plan in a completely feedforward fashion: it would require all motor parameters to be specified a priori, and thus would require exact knowledge of the position and properties of all necessary equipment and ingredients, such as weight, friction coefficients, etc.

Instead, many of these parameters can be filled in online by using information from the environment. It is not necessary to know the exact weight of a pan, because you can determine that easily by picking it up: you increase the exerted force until the pan leaves the surface of the kitchen counter. Although, you likely also learn a distribution of probable pan weights (e.g., more than 50 g and less than 10 kg) from your experience of other pans–or even just similarly-sized objects.

Interaction between feedforward and feedback becomes even more apparent on a higher level when a planned action fails to be executed. When a necessary ingredient is missing, replanning (or cancellation) of a preprogrammed action sequence may be necessary: if there is no butter, can I use oil to grease up the pan? Somehow, this information gathered by feedback processes must be communicated to the higher level action planner.

9.2.3   Feedforward and Feedback Control in Robots

The theorizing on action control in robotic systems must be considered rather ideological, sometimes driven by the specifics of particular robots and/or tasks considered and sometimes by broadly generalized anti-representationalist attitudes. Many early robots only had a handful of sensors and responded in a fixed pattern of behavior given a particular set of stimuli. Some robots were even purely feedforward, performing the same action or action sequence, with no sensory input whatsoever [37]. Feedforward or simple reactive control architectures make for very brittle behavior: even complex, carefully-crafted sequences of actions and reactions will appear clumsy if the environment suddenly presents an even slightly novel situation.

More complex architectures have been proposed, often with some analogy to biology or behavior, giving birth to the field of *behavior-based robotics*. The *subsumption architecture* [9] was a response to the traditional GOFAI, and posited that complex behavior need not necessarily require a complex control system. Different behaviors are represented as layers that can be inhibited by other layers. For example, a simple robot could be provided with the behaviors *wandering, avoiding, pickup,* and *homing.* These behaviors are hierarchically structured, with each behavior inhibiting its preceding behavior [4].

This hierarchy of inhibition between behavior is (although somewhat more complex) also visible in humans. For example, if your pants are (accidentally) set on fire while doing the dishes, few people would finish the dishes before stopping, dropping, and rolling. In other words, some behaviors take precedence over others. An approach similar to the subsumption architecture has been proposed by [3]. The *motor schema* approach also uses different, parallel layers of behavior, but does not have the hierarchical coordination as the subsumption approach does. Instead, each behavior contributes to the robot's overall response.

On a higher level, as noted in the previous section, other problems arise. When a planned action fails to succeed, for example because a robot can't find a pan to make pancakes in, replanning is necessary. The earliest AI planners such as GPS would simply backtrack to the previous choice point and try an alternative subaction. However, this does not guarantee the eventual successful completion of the action. Other planners, such as ABSTRIPS [41], use a hierarchy of representational levels. When it fails to complete a subaction, it could return to a more abstract level.

However, truly intelligent systems should be more flexible in handling such unforeseen events. If a robot cannot make me a pizza with ham, maybe it should make me one with bacon? Generalizing and substituting appropriate remain an elusive ability for robots, although vector space models of semantics (e.g., BEAGLE; [28]) offer a step in the right direction. Like neural networks, these models represent items (e.g., words) in a distributed fashion, using many-featured vectors with initially low similarity between random items. As the model learns–say, by reading documents, item representations are updated to make them more similar (on a continuous scale) to contextually similar items. These continually-updated representations can be used to extract semantic as well as syntagmatic (e.g., part-of-speech) relationships between items. Beyond text learning, vector space models may ultimately be used to learn generalizable representations for physical properties and manipulations of objects and environments.

9.2.4   Robotic Action Planning

It is understood that reaching movements in humans have an initial ballistic, feedforward component, followed by a slower, feedback-driven component that corrects for error in the initial movement. As people become more adept at reaching to targets at particular distances, a greater portion of their movement is devoted to the initial feedforward component, and less time is spent in the feedback component, thus speeding response times. Understanding how this happens should enable roboticists to make more adaptive, human-like motor planning systems for robots.

In this line of research, Kachergis et al. [29] studied sequence learning using mouse movements. Inspired by earlier work of Nissen and Bullemer [38], subsequences of longer sequences were acquired by human participants during a learning phase. The participants seem to implicitly extract the subsequences from longer sequences by showing faster response times and context effects.

These findings cast doubt on a simple chaining theory of sequential action. Rosenbaum et al. [39] interpreted these findings as evidence that sensory feedback is not a necessary component for action sequencing, in keeping with the conclusion of Lashley [34]. They argued that "the state of the nervous system can predispose the actor to behave in particular ways in the future," (p. 526), or, there are action plans for some behaviors. And yet, studies on spontaneous speech repair (e.g., [36]) also show that people are very fast in fixing errors in early components of a word or sentence, much too fast to assume that action outcomes are evaluated only after entire sequences are completed. This means that action planning cannot be exclusively feedforward, as Lashley [34] seemed to suggest, but must include several layers of processing, with lower levels continuously checking whether the current action component proceeds as expected. In other words, action planning must be a temporally extended process in which abstract representations to some extent provide abstract goal descriptions, which must be integrated with lower-level subsymbolic representations controlling sensorimotor loops. The existence of subsymbolic sensorimotor representations would account for context and anticipation effects, as described above.

The main lesson for robotic motor planning is that purely symbolic planning may be too crude and context-insensitive to allow for smooth and efficient multi-component actions. Introducing multiple levels of action planning and action control may complicate the engineering considerably, but it is also likely to make robot action more flexible and robust–and less "robotic" to the eye of the user.

9.3 ACQUISITION OF ACTION CONTROL

9.3.1 Introduction

In order for humans or robots to be able to achieve their goals, it is necessary for them to know what effect an action would have on their environment. Or, reasoning back, what actions are required to produce a certain effect in the environment. Learning relevant action-effect bindings as an infant is a fundamental part of development and likely bootstraps later acquisition of general knowledge.

In humans, learned action-effects seem to be stored bidirectionally. Following Lotze [35] and Harless [21], James [27] noted that intentionally creating a desired effect requires knowledge about, and thus the previous acquisition of action-effect contingencies. The *Theory of Event Coding* (TEC; [25]) is a comprehensive empirically well-supported (for recent reviews, see [23, 45]) theoretical framework explaining the acquisition and use of such action-effect bindings for goal-directed action. TEC states that actions and their expected effects share a common neural representation. Therefore, performing an action activates the expectation of relevant effects and thinking of (i.e., intending or anticipating) an action's effects activates motor neurons responsible for achieving those effects.

9.3.2 Human Action–Effect Learning

9.3.2.1 Traditional Action–Effect Learning Research

In traditional cognitive psychology experiments, action-effect bindings are acquired by having humans repetitively perform an action (such as pressing a specific button on a keyboard), after which an effect (such as a sound or a visual stimulus) is presented. After a certain amount of exposure to this combination of action and effect, evidence suggests that a bidirectional binding has been formed. When primed with a previously learned effect, people respond faster with the associated action [15]. This action-effect learning is quite robust but sensitive to action-effect contingency and contiguity [16].

9.3.2.2 Motor Babbling

Of course, action-effect learning does not only happen in artificial environments such as psychology labs. In fact, action-effect learning in humans starts almost instantly after birth [52] and some would argue even before. Young infants perform uncoordinated movements known as *body* or *motor babbling*. Most of these movements will turn out to be useless, however, some of them will have an effect that provides the infant with positive feedback. For example, a baby could accidentally push down with its right arm while lying on its belly, resulting in rolling on its back and seeing all sorts of interesting things. Over time, the infant will build up action-effect associations for actions it deems useful, and can perform motor acts by imagining their intended effects.

Having mastered the intricacies of controlling the own body, higher level action-effects can be learned in a manner similar to motor babbling. Eenshuistra et al. [14] give the example of driving a spacecraft that you are trying to slow down. If nobody ever instructed you on how to do that, your best option would probably be pressing random buttons until the desired effect is reached (be careful with that self-destruct button!). Once you have learned this action-effect binding, performance in a similar situation in the future will be much better.

9.3.3 Robotic Action–Effect Learning

The possibility that cognition can be grounded in sensorimotor experience and represented by automatically created action-effect bindings has attracted some interest of cognitive roboticists already. For instance, Kraft et al. [32] have suggested a three-level cognitive architecture that relies on object-action complexes, that is, sensorimotor units on which higher-level cognition is based. Indeed, action-effect learning might provide the cognitive machinery to generate action-guiding predictions and the off-line, feedforward component of action control. This component might specify the invariant aspects of an action, that is, those characteristics that need to be given for an action to reach its goal, to create its intended effect while an online component might provide fresh environmental information to specify the less goal-relevant parameters, such as the speed of a reaching movement when taking a sip of water from a bottle [24]. Arguably, such a system would have the benefit of allowing for more interesting cognitive achievements than the purely online, feedback-driven systems that are motivated by the situated-cognition approach [10]. At the same time, it would be more flexible than systems that rely entirely on the use of internal forward models [13]. Thus, instead of programmers trying to imagine all possible scenarios and enumerate reasonable responses, it might be easier to create robots that can learn action-effect associations appropriate to their environment and combine them with online information.

In robots as well as in humans, knowledge about one's own body is required to acquire knowledge about the external world. Learning how to control your limbs–first separately and then jointly (e.g., walking)–clearly takes more than even the first few years of life: after learning to roll over, crawl, and then walk, we are still clumsy at running and sport for several years (if, indeed, we ever become very proficient). Motor babbling helps develop tactile and proprioception–as well as visual and even auditory cues–of what our body in motion feels like. Knowing these basic actions and their effects on ourselves (e.g., what hurts) lays the foundation for learning how our actions can affect our environments.

In perhaps the first ever study of motor babbling in a (virtual) robot, Kuperstein [33] showed how random movement execution can form associations between a perceived object-in-hand position and the corresponding arm posture. This association is bidirectional, and as such is in line with ideomotor (or TEC) theory. We (and others, e.g., [11]) believe that such bidirectional bindings can help robots overcome traditional problems, such as inverse model inference from a forward model.

More recent investigations in robotic motor babbling have extended and optimized the method to include behavior that we would consider *curiosity* in humans. For example, Saegusa et al. [42] robotically implemented a sensorimotor learning algorithm that organized learning in two phases: *exploration* and *learning*. In the exploration stage, random movements are produced, while in the learning stage the action-effect bindings (or, more specifically, mapping functions) are optimized. The robot can then decide to learn bindings that have not yet been learned well.

9.4 DIRECTIONS FOR THE FUTURE

9.4.1 Introduction

Many questions remain with respect to the acquisition and skillful performance of not only well-specified, simple actions (e.g., reaching to a target) but of complex actions consisting of various components and involving various effectors. Indeed, how can we create a learning algorithm that can go from basic motor babbling to both successful goal-directed reaching, grasping, and manipulations of objects? To accomplish this obviously difficult goal, it will likely be beneficial for psychologists to study infants' development of these abilities and beneficial for cognitive roboticists to learn more from human capabilities.

9.4.2 Affordance Learning

Object manipulation and use is an indispensable activity for robots working in human environments. Perceiving object affordances–i.e., what a tool can do for you or how you can use an object–seems to be a quick, effortless judgment for humans, in many cases. For example, when walking around and seeing a door, you automatically pull the handle to open it.

One of the ways robots can perform object affordance learning is by motor babbling using simple objects as manipulators (e.g., [47]). In a so-called *behavioral babbling stage* a robot applies randomly chosen behaviors to a tool and observes their effects on an object in the environment. Over time, knowledge about the functionality of a tool is acquired, and can be used to manipulate a novel object with the tool.

As impressive as this may sound, this approach does not allow for easy generalization, and the robot cannot use this knowledge to manipulate objects using another, similar, tool. More recent approaches, such as demonstrated by Jain and Inamura [26] infer functional features from objects to generalize affordances to unknown objects. These functional features are supposed to be object invariant within a tool category.

In humans, an approach that seems successful in explaining affordance inference is based on Biederman's *recognition-by-components theory* [6]. This theory allows for object recognition by segmenting an encountered object in elementary geometric parts called *geons*. These are simple geometric shapes such as cones, cylinders and blocks. By reducing objects to a combination of more elementary units invariance is increased, simplifying object classification. Biederman recognized 36 independent geons, having a (restricted) generative power of 154 million three-geon objects.

In addition to being useful for object classification, geons can also be used to infer affordances. For example, a spoon is suitable for scooping because its truncated hollow sphere at the end of its long cylinder allows for containing things, and an elongated cylinder attached to an object can be used to pick it up.

One very promising example of the use of geons in affordance inference is demonstrated by Tenorth and Beetz [50]. This technique matches perceived objects to three-dimensional CAD models from a public database such as Google Warehouse. These models are then segmented into geons, which makes affordance inference possible.

However, the affordances that geons give us need to be learned in some way. Teaching robots how to infer what a tool can be capable of remains difficult. Ultimately, we want affordances to develop naturally during learning: be it from watching others, from verbal instruction, or from embodied experimentation. Task context is also an important aspect of affordance learning: depending on the situation, a hammer can be used as a lever, a paperweight, a missile, or well, a hammer. To understand how context affects action planning, studying naturalistic scenes and human activities jointly seems essential (cf. [2]).

Learning geon affordances that can be generalized to object affordances seems a fruitful approach to automating affordance learning in robots, although it is early to say whether this or other recent approaches will fare better. For example, deep neural networks use their multiple hidden layers along with techniques to avoid overfitting to learn high-level perceptual features for discriminating objects. The representations learned by such networks are somewhat more biologically-plausible than geon decompositions, and thus may be more suitable for generalization (although cf. [48] for generalization problems with deep neural networks).

9.4.3 Everyday Action Planning

A major obstacle in the way of robots performing everyday actions is the translation of high-level, symbolic task descriptions into sensorimotor action plans. In order to make such translations, one method would be to learn the other way around: by observing sensorimotor actions, segment and classify the input.

Everyday action is characterized by sequential, hierarchical action subsequences. Coffee- and tea-making tasks, for example, have shared subsequences such as adding milk or sugar. Moreover, the goal of adding sugar might be accomplished in one of several ways: e.g., tearing open and adding from a packet, or spooning from a bowl or box. Also, these subsequences do not necessarily have to performed in the same order every time (with some constraints, of course). It is this flexibility and ability to improvise that makes everyday action so natural for humans, yet so hard for robots.

Cognitive models that represent hierarchical information have been proposed (e.g. [12], [7]), but differ in the way they represent these hierarchies. One approach explicitly represents action hierarchies by hard-coding them into the model–hardly something we can do for a general autonomous robot– whereas the latter models hierarchy as an emergent property of the recurrent neural network. More recently, the model put forth by Kachergis et al. [30], uses a neural network with biologically plausible learning rules to extract hierarchies from observed sequences, needing far fewer exemplars than previous models.

9.5 CONCLUSION

In this chapter, we have discussed several concepts that are shared between cognitive robotics and cognitive psychology in order to argue that the creation of flexible, truly autonomous robots depends on the implementation of algorithms that are designed to mimic human learning and planning. Thus, there are many relevant lessons from cognitive psychology for aspiring cognitive roboticists.

Ideomotor theory and its implementations such as TEC provide elegant solutions to action-effect learning. Robotic motor learning algorithms that use motor babbling to bootstrap higher-order learning seem to be promising, and require little *a priori* knowledge given by the programmer, ultimately leading to more flexible robots.

Generalization of action plans is still a very difficult problem. Inferring hierarchical structure of observed or learned action sequences seems to be a promising approach, although the structure of everyday action seems to be nearly as nuanced and intricate to untangle as the structure of human natural language–and less well-studied, at this point. Again, we believe that biologically inspired learning models such as LeabraTI can play a role in making robotic action more human-like.

The overlapping interests of cognitive robotics and cognitive psychology has proven fruitful so far. Mechanisms like motor babbling and affordance inference, which are extensively studied in humans, can provide robots with techniques to make their behavior more flexible and human-like. We believe human inspiration for robots can be found at an even lower level by incorporating biologically-inspired neural models for learning in robots.

9.6 Bibliography

[1] J. A. Adams. A closed-loop of motor learning. *Journal of Motor Behavior*, 3:111–150, 1971.

[2] E. E. Aksoy, B. Dellen, M. Tamosiunaite, and F. Worgotter. Execution of a dual-object (pushing) action with semantic event chains. In *11th IEEE-RAS International Conference on Humanoid Robots (Humanoids)*, 2011.

[3] R. C. Arkin. Motor schema-based mobile robot navigation. *International Journal of Robotics Research*, 8:92–112, 1989.

[4] R. C. Arkin. Behavior-based robotics. MIT Press: Cambridge, MA, 1998.

[5] W. R. Ashby. *An introduction to cybernetics*. London: Chapman & Hall, 1956.

[6] I. Biederman. Recognition-by-components: A theory of human image understanding. *Psychological Review*, 94:115–147, 1987.

[7] M. Botvinick and D. C. Plaut. Doing without schema hierarchies: a recurrent connectionist approach to normal and impaired routine sequential action. *Psychological Review*, 111(2):395–429, Apr 2004.

[8] V. Braitenberg. *Vehicles: Experiments in synthetic psychology*. Cambridge, MA: MIT Press, 1984.

[9] R. Brooks. A robust layered control system for a mobile robot. *IEEE Journal of Robotics and Automation*, 2:14–23, 1986.

[10] R. Brooks. Intelligence without reason. In *Proceedings of the 12th International Joint Conference on Artificial Intelligence*, volume 1, pages 569–595, 1991.

[11] D. Caligiore, D. Parisi, N. Accornero, M. Capozza, and G. Baldassarre. Using motor babbling and Hebb rules for modeling the development of reaching with obstacles and grasping. In *Proceedings of the 2008 International Conference on Cognitive Systems*, 2008.

[12] R. P. Cooper and T. Shallice. Hierarchical schemas and goals in the control of sequential behavior. *Psychological Review*, 113(4):887–916; discussion 917–31, Oct 2006.

[13] Y. Demiris and A. Dearden. From motor babbling to hierarchical learning by imitation: a robot developmental pathway. In *International Workshop on Epigenetic Robotics*, pages 31–37, 2005.

[14] R.M. Eenshuistra, M.A. Weidema, and B. Hommel. Development of the acquisition and control of action-effect associations. *Acta Psychologica*, 115:185–209, 2004.

[15] B. Elsner and B. Hommel. Effect anticipation and action control. *Journal of Experimental Psychology: Human Perception and Performance*, 27:229–240, 2001.

[16] B. Elsner and B. Hommel. Contiguity and contingency in the acquisition of action effects. *Psychological Research*, 68:138–154, 2004.

[17] R. Fikes and N. Nilsson. Strips: a new approach to the application of theorem proving to problem solving. *Artificial Intelligence*, 2:189–208, 1971.

[18] S. Glover. Separate visual representations in the planning and control of action. *Behavioral and Brain Sciences*, 27:3–24, 2004.

[19] M. A. Goodale and A. D. Milner. Separate visual pathways for perception and action. *Trends in Neurosciences*, 15:20–25, 1992.

[20] M. A. Goodale, D. Pelisson, and C. Prablanc. Large adjustments in visually guided reaching do not depend on vision of the hand or perception of target displacement. *Nature*, 320:748–750, 04 1986.

[21] E. Harless. Der Apparat des Willens. *Zeitschrift für Philosophie und philosophische Kritik*, 38:50–73, 1861.

[22] F. M. Henry and D. E. Rogers. Increased response latency for complicated movements and a "memory drum" theory of neuromotor reaction. *Research Quarterly*, 31:448–458, 1960.

[23] B. Hommel. Action control according to TEC (theory of event coding). *Psychological Research*, 73:512–526, 2009.

[24] B. Hommel. *Effortless attention: A new perspective in the cognitive science of attention and action*, chapter Grounding attention in action control: The intentional control of selection, pages 121–140. Cambridge, MA: MIT Press, 2010.

[25] B. Hommel, J. Müsseler, G. Aschersleben, and W. Prinz. The theory of event coding (TEC): A framework for perception and action planning. *Behavioral and Brain Sciences*, 24:849–937, 2001.

[26] R. Jain and T. Inamura. Bayesian learning of tool affordances based on generalization of functional feature to estimate effects of unseen tools. *Artificial Life and Robotics*, 18:95–103, 2013.

[27] W. James. *Principles of psychology*, volume 1. New York: Holt, 1890.

[28] M. N. Jones and D. J. K. Mewhort. Representing word meaning and order information in a composite holographic lexicon. *Psychological Review*, 114:1–37, 2007.

[29] G. Kachergis, F. Berends, R. de Kleijn, and B. Hommel. Trajectory effects in a novel serial reaction time task. In *Proceedings of the 36th Annual Conference of the Cognitive Science Society*, 2014.

[30] G. Kachergis, D. Wyatte, R. C. O'Reilly, R. de Kleijn, and B. Hommel. A continuous time neural model for sequential action. *Philosophical Transactions of the Royal Society B*, 369:20130623, 2014.

[31] H. D. Knapp, E. Taub, and A. J. Berman. Movements in monkeys with deafferented forelimbs. *Experimental Neurology*, 7:305–315, 1963.

[32] D. Kraft, E. Baseski, M. Popovic, A.M. Batog, A. Kjær-Nielsen, N. Krüger, R. Petrick, C. Geib, N. Pugeault, M. Steedman, T. Asfour, R. Dillmann, S. Kalkan, F. Wörgötter, B. Hommel, R. Detry, and J. Piater. Exploration and planning in a three level cognitive architecture. In *Proceedings of the International Conference on Cognitive Systems (CogSys 2008)*,, Karlsruhe, 2008.

[33] M. Kuperstein. Neural model of adaptive hand-eye coordination for single postures. *Science*, 239:1308–1311, 1988.

[34] K. S. Lashley. *Cerebral mechanisms in behavior*, chapter The problem of serial order in behavior, pages 112–131. New York: Wiley, 1951.

[35] R. H. Lotze. *Medicinische Psychologie oder Physiologie der Seele*. Weidmann, 1852.

[36] C. Nakatani and J. Hirschberg. A corpus-based study of repair cues in spontaneous speech. *Journal of the Acoustical Society of America*, 95:1603–1616, 1994.

[37] S. B. Niku. *Introduction to robotics: analysis, control, applications*. Wiley, 2010.

[38] M. J. Nissen and P. Bullemer. Attentional requirements of learning: Evidence from performance measures. *Cognitive Psychology*, 19:1–32, 1987.

[39] D. A. Rosenbaum, R. G. Cohen, S. A. Jax, D. J. Weiss, and R. van der Wel. The problem of serial order in behavior: Lashley's legacy. *Human Movement Science*, 26:525–554, 2007.

[40] D. E. Rumelhart and J. L. McClelland. *Parallel distributed processing: Explorations in the microstructure of cognition. Volume I.* Cambridge, MA: MIT Press, 1986.

[41] E. D. Sacerdoti. Planning in a hierarchy of abstraction spaces. *Artificial Intelligence*, 5:115–135, 1974.

[42] R. Saegusa, G. Metta, G. Sandini, and S. Sakka. Active motor babbling for sensory-motor learning. In *IEEE International Conference on Robotics and Biomimetics*, pages 794–799, 2008.

[43] R. A. Schmidt. A schema theory of discrete motor skill learning. *Psychological Review*, 82:225–260, 1975.

[44] R. D. Seidler, D. C. Noll, and G. Thiers. Feedforward and feedback processes in motor control. *Neuroimage*, 22:1775–1783, 2004.

[45] Y. K. Shin, R. W. Proctor, and E. J. Capaldi. A review of contemporary ideomotor theory. *Psychological Bulletin*, 136(943–974), 2010.

[46] B. F. Skinner. *The behavior of organisms: An experimental analysis*. Cambridge, Massachusetts: B.F. Skinner Foundation, 1938.

[47] A. Stoytchev. Autonomous learning of tool affordances by a robot. In *AAAI 2005*, 2005.

[48] C. Szegedy, W. Zaremba, I. Sutskever, J. Bruna, D. Erhan, I. J. Goodfellow, and R. Fergus. Intriguing properties of neural networks. *CoRR*, abs/1312.6199, 2013.

[49] E. Taub, I. A. Goldberg, and P. Taub. Deafferentation in monkeys: Pointing at a target without visual feedback. *Experimental Neurology*, 46:178–186, 1975.

[50] M. Tenorth and M. Beetz. KnowRob: A knowledge processing infrastructure for cognition-enabled robots. Part 1: The KnowRob system. *International Journal of Robotics Research*, 2013.

[51] A. M. Turing. Computing machinery and intelligence. *Mind*, 59:433–460, 1950.

[52] S. A. Verschoor, M. Spapé, S. Biro, and B. Hommel. From outcome prediction to action selection: Developmental change in the role of action-effect bindings. *Developmental Science*, 16:801–814, 2013.

[53] W. G. Walter. *The living brain*. London: Duckworth, 1953.

[54] J. B. Watson. Psychology as the behaviourist views it. *Psychological Review*, 20:158, 1913.

[55] R. S. Woodworth. The accuracy of voluntary movement. *Psychological Review*, 3:1–119, 1899.

VII

Artificial Intelligence Aspect of Cognitive Robotics

CHAPTER 10

A Bottom-Up Integration of Vision and Actions To Create Cognitive Humanoids

Jürgen Leitner

Dalle Molle Institute for AI (IDSIA) & SMRTRobots,
Lugano, Switzerland

CONTENTS

I N recent years more and more complex humanoid robots have been developed. On the other hand programming these systems has become more

difficult. There is a clear need for such robots to be able to adapt and perform certain tasks autonomously, or even learn by themselves how to act. An important issue to tackle is the closing of the sensorimotor loop. Especially when talking about humanoids the tight integration of perception with actions will allow for improved behaviours, embedding adaptation on the lower-level of the system.

10.1 INTRODUCTION

Object manipulation in real-world settings is a very hard problem in robotics, yet it is one of the most important skills for robots to possess [29]. Through manipulation they are able to interact with the world and therefore become useful and helpful to humans. Yet to produce even the simplest human-like behaviours, a humanoid robot must be able to see, act, and react continuously. Even more so for object manipulation tasks, which require precise and coordinated movements of the arm and hand. The understanding of how humans and animals control these movements is a fundamental research topic in cognitive- [49] and neuro-sciences [27]. Despite the interest and importance of the topic, e.g. in rehabilitation and medicine, the issues and theories behind how humans learn, adapt and perform reaching and grasping behaviours remain controversial. Although there are many experimental studies on how humans perform these actions, the development of reaching and grasping is still not fully understood and only very basic computational models exist [45]. Vision is seen as an important factor in the development of reaching and grasping skills in humans [5, 39]. For example, imitation of simple manipulation skills has been observed already in 14-month-old infants [40]. Current robots in contrast are only able to perform (simple) grasps in very limited, specific settings. To enable more autonomous object manipulation, more specifically how to enable some level of eye-hand coordination to perform actions more successfully, is of high interest to the robotics community (see e.g. NASA's Space Technology Roadmap calls for "Real-time self-calibrating hand-eye System" [1]).

Artificial Intelligence, Machine Learning and Robotics

The research in the fields of Artificial Intelligence (AI) and robotics were strongly connected in the early days, but have diverged over the last decades. Although AI techniques were developed to play chess on a level good enough to win against (and/or tutor) the average human player [51], the robotic manipulation of a chess piece, in contrast, the creation of intelligent machines has lacked quite a bit behind the algorithmic side. It is still not feasible to control a robot on a similar level of precision, adaptation and success as a human — not even comparative to children level. To produce even the simplest autonomous, adaptive, human-like behaviours, a humanoid robot must be able to, at least:

- Identify and localize objects in the environment, e.g. the chess pieces and board

- Execute purposeful motions for interaction, e.g. move a piece to a desired position

At the beginning of AI research a clear goal was to build complete, intelligent, autonomous robotic system [50]. As with the example of the above example of chess, it has proven to be quite challenging. Not helping the cause was the fractioning of the fields into many distinct facets of research. While there was progress in each of the sub-fields and the both disciplines (AI and robotics) separately, it has now become clear that a closer integration is again needed. There has been a renewed interest, from both research communities, to work together again towards the goal of intelligent robotic systems.

The field of robotics has clearly matured over the last few years. Current humanoid robots are stunning feats of engineering as mention above. To embed this systems with some sense of 'intelligence' and use the full versatility of advanced robotic systems, a bigger collaboration with the research community in Artificial Intelligence and Machine Learning is required.

The idea of the 'embodied mind' stems from philosophy. It claims that the nature of the human mind is determined by the form of the human body. Philosophers, psychologists, cognitive scientists, and artificial intelligence researchers who study embodied cognition and the embodied mind argue that all aspects of cognition are shaped by aspects of the body. The embodied mind thesis is opposed to other theories of cognition. Embodied cognition reflects the argument that the motor system influences our cognition, just as the mind influences bodily actions. Roboticists have argued that to understand intelligence and build artificial system that comprise intelligence can only be achieved by machines that have both sensory and motor skills. Furthermore they need to be interacting with the world through a body. This 'embodiment' is seen as an important condition for the development of cognitive abilities both in humans and robots [9, 62, 48]. The insights of these robotics researchers have in return also influenced philosophers.

Machine Learning algorithms, have been applied in experimental robotics to acquire new skills, however the need for carefully gathered training data, clever initialization conditions, and/or demonstrated example behaviours limits the autonomy with which behaviours can be learned. To build robots that can perform complex manipulation skills that help users in their activities of daily living is the aim of various research projects in Europe (e.g. [11, 60]).

Robot Learning

As mentioned above the programming of these highly complex robot systems is a cumbersome, difficult and time-consuming process. Current approaches tend to describe each precise each precise movement in detail, allowing little to no flexibility or adaptation during execution. This obviously has issues with scaling to highly complex robots in complicates settings. Therefore the robotics community has focused on methods to provide robots with the ability to act autonomously, adapt or 'learn' how to behave without the need of hard-coding every possible outcome.

Autonomous robots research is aimed at building systems that do not require the pre-programming of every possible situation encountered. Many kinds of robots have some degree of autonomy and different robots can be autonomous in different ways. In fields, such as space exploration, a high degree of autonomy is desirable. For an autonomous robot one generally assumes the following capabilities [14]:

- Gain information about the environment (Rule #1)

- Work for an extended period without human intervention (Rule #2)

- Move either all or part of itself throughout its operating environment without human assistance (Rule #3)

- Avoid situations that are harmful to people, property, or itself unless those are part of its design specifications (Rule #4)

- Maintain its own survival at the expense of the previous rules (Sentient Robot Mandate) (Rule #5)

- Learn or gain new capabilities like adjusting strategies for accomplishing its task(s) or adapting to changing surroundings (Rule #6)

In the early 90s of the last century *Behavioural Robotics* (or behaviour-based robotics) was introduced as a way to deal with more and more complex robots and application areas [8]. This research area focuses on flexible switching mechanisms to change the robots main behaviours based only on a very simple internal model. The basic idea is that close (and probably simple) sensor-motor connections can result in behaviours that appear complex and sophisticated. Due to the fact that these models used a simple approach, rather than a computational complex model and the relatively low cost of development, popularised this approach in the mid-1990s. This paradigm has had a wide range of application in multi-robot teams [4] yet the scaling to complex robots, such as humanoid, has not been successful so far.

Robot Learning generally refers to research into ways for a robot to learn certain aspects by itself. Instead of providing all information to the robot a priori, for example, possible motions to reach a certain target position, the agent will through some process 'learn' which motor commands lead to what action. The research field is placed at the intersection of machine learning and robotics and studies how robots can acquire new skills through experimentation. The earlier mentioned 'embodiment' plays an important role here. Example include the learning of sensorimotor skills (for example locomotion, grasping, object manipulation), as well as interactive skills such as manipulation of an object in collaboration with a human. In addition the learning of linguistic skills, especially the grounding of words or phrases in the real world, is of interest to the research community. The field of 'robot learning' is closely related to other disciplines, for example, adaptive control. Learning in realistic environments requires algorithms that can deal with high-dimensional states, e.g. to detect events in the stream of sensory inputs, change and uncertainty. Note that while machine learning is nowadays often used for computer and robot vision tasks (like in this dissertation), this area of research are usually not referred to as 'robot learning'. The fields of *Cognitive Robotics*, *Developmental Robotics* and *Evolutionary Robotics* emerged with the specific aim to investigate how robots can 'learn' for themselves and thereby generate more autonomous and adaptive capabilities.

In *Cognitive Robotics* [3] the aim is to provide robots with cognitive processes, similar to humans and animals. An integrated view of the body is taken, including the motor system, the perceptual system and the body's interactions with the environment. The acquisition of knowledge, may it be through actions (e.g. motor babbling) or perception is a big part of cognitive robotics research. Another is the development of architectures for these tasks. A variety has been proposed [54, 58, 12, 63], but the promised improvements in robotic applications still need to be shown. This can be attributed to the varying definitions of cognition and the complex human cognitive system, whose workings are still not fully understood. To build cognitive architectures two distinct approaches have been tried. The research seems to mainly focus on top-down architectures. A bottom-up approach has been described as suitable for the use with robots (e.g. the proposed *iCub* cognitive architecture [59]).

Developmental Robotics [3, 61, 2] is aiming to put more emphasis on the development of skills. It is an interdisciplinary approach to developmental science. It differs from the previous approaches, as the engineer only creates the architecture and then allows the robot to explore and learn its own representation of its capabilities (sensory and motor) and the environment. As above, the body and its interactions with the environment are seen as being fundamental for the development of skills. Aims are to build adaptive robotic systems by exploration and autonomous learning, i.e. learning without a direct intervention from a designer [37]. Here interesting areas to explore are selected by building on previous knowledge, while seeking out novel stimuli.

Evolutionary Robotics [23, 43] is another approach to add adaptiveness and developmental processes to robots. It emerged as a new approach to overcome the difficulties of designing control systems for autonomous robots: (a) coordinating the (increasing) number of DOF both in mechanics and control is hard, especially since the complexity scales with the number of possible interactions between parts (see 'Curse of Dimensionality' [15]) (b) the environment and how the robot interacts with it are often not known before. Its main focus is on evolve a control system based on artificial neural networks. These neuro-controllers (NC), inspired by the neuron activity in the human brain, have been shown to work in a wide range of applications [44, 17, 31]. An important issue is that to 'learn' behaviours, a large number of iterations (or generations) is required. This works fine in simulation but is hard to achieve on a real robotic platform. [44] showed that evolving a NC on hardware is, while time consuming, feasible, at least for simple mobile robots. Hybrid approaches, where NCs are trained first in simulation and then transferred to the real hardware, seem preferential. The performance of the controllers in the real world can then be used to improve the simulation [6]. How to effectively train and apply NCs to real, high-DOF hardware is still an open research question.

Other Approaches to robot learning have been developed in the past. The area of Reinforcement Learning (RL) [56] has appealed to many roboticists, especially for learning to control complex robotic systems. A general RL algorithm and the means to inform the robot whether its actions were successful (positive reward) or not (negative reward) is all that is required. RL and its applicability to humanoid robots has been investigated by [47]. Imitation Learning or Apprenticeship Learning is of importance in human skill development as it allows to transfer skills from one person to another. In robotics Robot Learning from demonstration or Programming by Demonstration is a similar paradigm for enabling robots to learn to perform novel tasks. It takes the view that an appropriate robot controller can be derived from observations of a another agent's performance thereof [53].

10.2 A COGNITIVE ROBOTICS APPROACH

One of the most important problems in robotics currently is arguably to improve the robots' abilities to understand and interact with the environment around them: a robot needs to be able to perceive, detect and locate objects in its surrounding and then then have the ability to plan and execute actions to manipulate these objects detected.

The described approach herein was developed to extend the capabilities of the *iCub* humanoid robot, especially to allow for more autonomous and more adaptive – some would say, more 'intelligent' – behaviours. The *iCub* is a state-of-the-art, high degree-of-freedom (DOF) humanoid (see Figure 10.1) [57]. It consists of two arms and a head attached to a torso roughly the size of a human child. The head and arms follow an anthropomorphic design and provide a high DOF system that was designed to investigate human-like object manipulation. It provides also a tool to investigate human cognitive and sensorimotor development. To allow for safe and 'intelligent' behaviours the robot's movements need to be coordinated closely with feedback from its sensors. The *iCub* is an excellent experimental platform for cognitive, developmental robotics and embodied artificial intelligence [42].

The aim is to generate a not before seen level of eye-hand coordination on the *iCub*. Pick-and-place operations were chosen as they require intelligent behaviour in a complex environment, i.e. perceiving which objects are in its vicinity, reaching for a specific object, while avoiding obstacles. The cognitive

FIGURE 10.1 The experimental platform used: the *iCub* humanoid robot.

skills, from learning what an object is and how to detect it in the sensory stream, to adapting the reach if the environment changes, are embedded using a variety of frameworks. First functional motion and vision subsystems are developed, which are then integrated to create a closed action-perception loop. The vision side detects and localises the object continuously, while the motor-side tries to reach for target objects avoiding obstacles at the same time.

A combination of robot learning approaches with computer vision and actions is used to improve adaptivity and autonomy in robot grasping based on visual feedback. The next section contains the description of the robot vision frameworks and techniques developed for and implemented on the *iCub*, shown in the top row of Figure 10.2 (in green). It includes the modules for the detection and identification of objects (in the images), as well as, the localization (in 3D Cartesian space). The bottom half, in yellow, shows the action and motion side. To generate motion using machine learning techniques a crucial feature is avoiding collisions, both between the robot and the environment *and* the robot and itself.

The various modules developed and interacting in this cognitive approach are the following:

- *Object Models, Detection and Identification:* as mentioned above, the detection and identification of objects is a hard problem. To perform these tasks CGP-IP (Cartesian Genetic Programming for Image Processing) [22] is used. It provides a machine learning approach to building visual object models, which can be converted into executable code, in both supervised and unsupervised fashion [32]. The resulting program performs the segmentation of the camera images for a specific object.

- *Object Localisation:* by using the image coordinates of the detected object from the two cameras together with the current robot's pose, the posi-

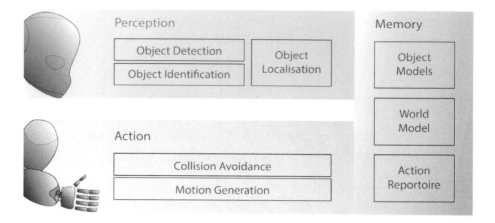

FIGURE 10.2 Overview of the proposed architecture for a functional eye-hand coordination on the *iCub* humanoid. The *object detection* and *identification* is currently solely based on the camera images (2D) received. The *object localization* uses the information from the two cameras to calculate an operational space (3D) position. This is the same space in which *collision avoidance* is applied and the *world* is modelled. The *object model* contains the information of how to detect the object in the 2D images (see Section 10.3). The *motion generation* and *action repertoire* can use the full configuration space of the humanoid (41 DOF).

tion of the object can be estimated in Cartesian space wrt. the robot's reference frame. Instead of a calibration for each single camera, the stereo system and the kinematic chain, a module that learns to predict these from a training set is incorporated. The system has been shown to estimate these positions with a technique based on genetic programming [35] and an artificial neural network estimators [34]. After the object is detected in the camera images the location of an object is estimated and the world model is updated.

- *Action Repertoire:* a light-weight, easy-to-use, one-shot grasping system (LEOGrasper[1]), which has been used extensively at IDSIA (Figure 10.6), provides the main grasping subsystem. It can be configured to perform a variety of grasps, all requiring to close the fingers in a coordinated fashion. A variety of more complex actions/roadmaps can be generated offline and later executed on the *iCub* [55], to e.g. lead to improved perception [36].

---

[1]Source code available at: https://github.com/Juxi/iCub/

- *World Model, Collision Avoidance and Motion Generation:* the world model keeps track of the robot's pose in space and the objects it has visually detected. Figure 10.8 shows this model including the robot's pose the static table, and the two objects localised from vision. MoBeE is used to safeguard the robot from (self-)collisions. It furthermore allows to generate motion by forcing the hand in operational space.

Section 10.4 describes the developed techniques to control the *iCub*. All these subsystems are supported by memory (in blue) enabling the persistent modelling of the world and providing a repertoire of actions to be triggered. In Section 10.5 the tight integration of these two sides, the perception and the motion, is described. Section 10.6 presents a proof-of-concept, highlighting a level of eye-hand coordination not previously seen on the *iCub*.

10.3 PERCEIVING THE ENVIRONMENT

To be useful in the above proposed scenarios a robot must be able to see, act, and react continuously. Perception is a key requirement in order to purposefully adapt robot motion to the environment, allowing for more successful, more autonomous interactions. The first important step towards this is to understand the environment the robot is embedded in. Coming back to the example of playing chess, this would compare to finding the chess board and each of the chess pieces (e.g. in a camera image) or even just to to realise that there is a chess board and pieces in the scene.

Vision and the visual system are the focus of much research in psychology, cognitive science, neuroscience and biology. A major problem in visual perception is that what individuals 'see' is not just a simple translation of input stimuli (compare *optical illusions*). One important area of research to build robots that can understand their surroundings is the development of artificial vision. *Computer Vision* – sometimes referred to as *Robot Vision* when applied in a robotic system – generally describes the field of research dealing with acquiring, processing, analysing, and understanding images in order to produce decisions based on the observation. The fields of computer vision and AI have close ties, e.g. autonomous planning or decision making for robots requires information about the environment, which could be provided by a computer vision system. AI and computer vision share other topics such as pattern recognition and learning techniques.

Even though no clear definition of the areas of computer vision and image processing exists, the latter is commonly used to refer to a subsection of computer vision. Image processing techniques generally provide ways of extracting information from the image data, for example, noise reduction, feature extraction, segmentation, etc.[21] Another important topic in computer vision is 'image understanding'. With the aid of geometry, statistics, and learning the goal is to mimic the abilities of the human (visual) perception system.

Research into vision for the special requirements of robotic systems is referred to as *robot vision* or *machine vision* [24, 25]. For example, visual feedback has extensively been used in mobile robot applications, for obstacle avoidance, mapping and localization. With the advancement of humanoids and the increased interest in working around humans, object detection and manipulation are more and more driving the development of robot vision systems. An important problem is that of determining whether or not the image data contains some specific object, feature, or activity. While this has been researched for quite some time already, the task seems harder than expected and no solution for the general case of detecting arbitrary objects in arbitrary situations exists. From a robot vision point of view, this means that the robot is required to detect previously unknown objects in its surroundings and be able to build models to memorise and identify them in the future. Most of the work is heavily relying on artificial landmarks and fiducial markers to simplify the detection problem. Furthermore existing methods can at best solve it for specific objects (simple geometries, faces, printed or hand-written characters, or vehicles) and in specific situations (in terms of well-defined illumination, background, and pose of the object wrt. the camera). For a detailed introduction and overview of the foundations and the current trends the reader is referred to the excellent survey by [30].

10.3.1 Object Detection: icVision & CGP-IP

Aiming at eye-hand coordination and object manipulation the following frameworks were implemented to enable the learning and real-time operation of object detection and identification.

icVision [33] is an open-source[2], biologically-inspired framework consisting of distributed YARP [41] modules performing computer vision related tasks in support of cognitive robotics research (Figure 10.3). It includes the modules for the detection and identification of objects (in the camera images, referred to as *Filters*), as well as the localisation of the objects in the robot's operational space (3D Cartesian space). At the centre is the *icVision* core module, which handles the connection with the hardware and provides housekeeping functionality (e.g., extra information about the modules started and methods to stop them). Currently available modules include object detection, 3D localisation, gazing control (attention mechanism) and saliency maps. Standardised interfaces allow for easy swapping and reuse of modules.

The main part in object detection, the binary segmentation of the object from the background (see Figure 10.3 on the right), in the visual space, is performed in separate *icVision* filter modules. Each one is trained using the Cartesian Genetic Programming for Image Processing (CGP-IP) framework [22], in combination with the OpenCV [7] library, to detect and identify specific objects in a variety of real-life situations (e.g. a tea box as shown in Figure 10.4). The framework can run multiple filter modules in parallel. A

---

[2]Code available at: https://github.com/Juxi/icVision/

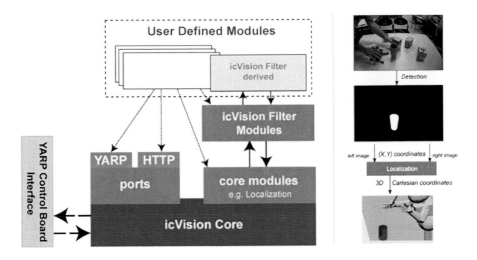

FIGURE 10.3 The icVision Architecture: The core module (bottom) is mainly for housekeeping and accessing & distributing the robot's camera images and motor positions. Object detection is performed in the *filter* modules (top), by segmenting the object of interest from the background. A typical work flow is shown (right): input images are retrieved from the cameras, the specific object is detected by a trained *Filter Module*, before the outputs (together with the robot's current pose) are used to estimate a 3D position using a localisation module. The object is then placed in the world model.

variety of filters have been learnt and most of them are able to perform the object detection in near real-time.

To interact with the objects the robot also needs to know where the object is located. Developing an approach to perform robust localisation to be deployed on a real humanoid robot is necessary to provide the necessary inputs for on-line motion planning, reaching, and object manipulation. *icVision* provides modules to estimate the 3D position based on the robot's pose and the location of object in the camera images.

10.4 INTERACTING WITH THE ENVIRONMENT

Computer vision has become a more and more prominent topic of research over the past decades, also in the field of robotics. Like humans and animals, robots are able to interact with the world around them. While most robot vision research tends focus on understanding the world from just passive observations, these interactions with the environment provide and create valuable information to build better visual systems. Connecting manipulation commands with visual inputs allows for a robot to create methods to actively explore its surroundings. These connections between motor actions and observations exist in the human brain and are an important aspect of human development [5].

Only after the scene is observed and the robot has an idea about which objects are in the environment, can it start interacting with these in a safe fashion. In the chess example, even if the state of the board and where it is located are known, to move a certain chess piece from one field to another without toppling other pieces is still a hard problem by itself. In fact, children even at a very young age, have significantly better (smoother, more 'natural', 'fluent' and controlled) hand movements than all currently available humanoid robots. But manipulating arbitrary objects is not a trivial thing, even for humans. The development of hand control in children, for an apparently simple, prototypical precision grasp task is not matured until the age of 8-10 years [18]. Moreover, complexity, as can be seen by the number of neurons comprising the control of the arm and hand, is staggeringly high. Even after manipulation skills have been learnt they are constantly adapted by an perception-action loop to yield desired results. In infants various specializations in the visual pathways may develop for extracting and encoding information relevant for visual cognition, as well as, information about the lo-

FIGURE 10.4 Detection of complex objects, e.g. a tea box, in changing poses, different light and when partially occluded is a hard problem in robot vision.

cation and graspability of objects [28]. This hints at the very close integration of vision and action in the human brain.

In recent years good progress was made with robotic grasping of objects. The various manipulators, mainly hands and grippers, and techniques clearly improved. Also novel concepts of 'grippers' have been designed and some are quite ingenious solutions to a number of issues. One such example is the granular gripper made by [10], which is made out of grounded coffee beans which are able to 'flow' around the object and then fixed in position by creating a vacuum. This concept has recently been extended to a full sized elephant-trunk-style arm [13]. Also in terms of how to grasp objects with regular grippers and 'hands' recent results highlight the advanced state of research in grasping. For example, [38], with their research showed that robots are able to pick up non-rigid objects, such as, towels. Their robot is able to reliably and robustly pick up a randomly dropped towel from a table by going through a sequence of vision-based re-grasps and manipulations-partially in the air, partially on the table. On the other hand, ways for a robot to learn from only a small number of real world examples, where good grasping points are on a wide variety of previously unknown objects have been presented [52].

The progress on performing grasping operations in the last years shows that one can now use these grasping subroutines and further integrate them in autonomous systems. The direct interface between various components, which makes robotics such a hard but interesting field, clearly needs to improve to allow for robust object manipulation. Only by combining sensing and control of the whole robotic platform a fully functional 'pick-and-place' capable system will appear. To allow for a variety of objects to be picked up from various positions the robot needs to see, act and react within a control system in which these elements are tightly integrated.

10.4.1 Collision Avoidance and World Model: MoBeE

An important issue is to ensure the safe operation of our humanoid. Modular Behavioral Environment (MoBeE) [19] is a software infrastructure to realise complex, autonomous, adaptive and foremost safe robot behaviours. It acts as an intermediary between three loosely coupled types of modules: the Sensor, the Agent and the Controller. These correspond to abstract solutions to problems in Computer Vision, Motion Planning, and Feedback Control, respectively. An overview of the system is depicted in Figure 10.5. The framework is robot independent, and can exploit any device that controlled via YARP [41]. It also supports multiple interacting robots, and behavioural components are portable and reusable thanks to their weak coupling. MoBeE controls the robot constantly, according to the following second order dynamical system:

$$M\ddot{q}(t) + C\dot{q}(t) + K(q(t) - q^*) = \sum f_i(t) \qquad (1)$$

where $q(t) \in R^n$ is the vector function representing the robot's configuration, M, C, K are matrices containing mass, damping and spring constants

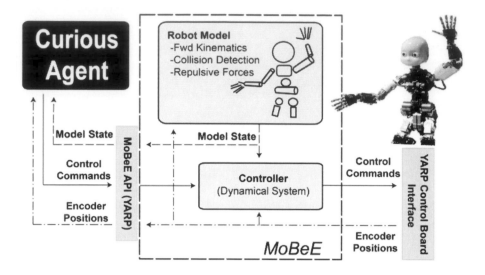

FIGURE 10.5 The Modular Behavioral Environment Architecture: MoBeE implements low-level control and enforces necessary constraints to keep the robot safe and operational in real-time. Agents (left) are able to send high-level commands, while a kinematic model (top) is driven by the stream of encoder positions (right). The model computes fictitious constraint forces, which repel the robot from collisions, joint limits, and other infeasibilities. These forces, $f_i(t)$, are passed to the controller (middle), which computes the attractor dynamics that governs the actual movement of the robot.

respectively. q^* denotes an attractor (resting pose) in configuration space. Constraints on the system are implemented by forcing the system via $f_i(t)$, providing automatic avoidance of kinematic infeasibilites arising from joint limits, cable lengths, and collisions.

An agent can interact with MoBeE, instead of directly with the robot, by sending arbitrary high-level control commands. For example, when a new attractor q^* is set to a desired pose by an agent, e.g. by calculating the inverse kinematics of an operational space point, $q(t)$ begins to move toward q^*. The action then terminates either when the dynamical system settles or when a timeout occurs, depending on the constraint forces $f_i(t)$ encountered during the transient response.

10.4.2 Action Repertoire: TRM & LEOGrasper

The action repertoire for the scenario herein consists mainly of a grasping subsystem and a framework to generate full-body motions. LEOGrasper is a

FIGURE 10.6 Grasping a variety of objects successfully, such as, tin cans, plastic cups and tea boxes. The module works for both the right and left hand.

light-weight, easy-to-use, one-shot grasping system (Figure 10.6).[3]), It can be configured to perform a variety of grasps, all requiring to close the fingers in a coordinated fashion. The *iCub* incorporates touch sensors on the fingertips, due to the high noise, we use the error reported by the PID controllers of the finger motors to know when they are in contact with the object. A variety of more complex actions/roadmaps can be generated offline and later executed on the *iCub* [55], to e.g. lead to improved perception [36].

To generate a set of more-complex actions to execute MoBeE's kinematic model was extracted and connected with a machine-learning based, black-box optimizer. The system aims to find find a robot pose $q_{goal} \in C$, where C describes the robot's configuration space, that satisfies some operational space constraints, with planning, i.e. find a feasible configuration-space trajectory, $Q \subset C$, which is the curve from the current pose, $q_{initial}$ and the target pose q_{goal}.

Natural Evolution Strategies (NES) are applies to find a set of task-related poses yielding Task-relevant Road Map (TRM) [55]. It finds a family of postures that are optimized under constraints defined by arbitrary cost-functions, and at the same time maximally covers a user-defined task-space. Connecting these postures creates a rather dense, traversable graph, which we call roadmaps. In other words, the task-relevant constraints are built directly into the TRM, and motion planning is reduced to graph search. This allows to build TRMs that can perform useful tasks in the 41-dimensional configuration space of the upper body of the *iCub* humanoid. Additionally these maps can be stored to create an action repertoire that can be recalled when a certain tasks needs to be executed. Figure 10.7 shows time-lapse snapshots of motions, planned within TRMs. It provides an idea of what kind of motions can be generated.

---

[3]Source code available at: https://github.com/Juxi/iCub/

10.5   INTEGRATION

To allow for a variety of objects to be picked up from various positions the robot needs to see, act and react within an integrated control system.

For example, methods enabling a 5 DOF robotic arm to pick up objects using a point-cloud generated model of the world and objects are available to calculate reach and grasp behaviours [52]. In 2010 a technique for robots to pick up non-rigid objects, such as, towels was presented [38]. It allows to reliably and robustly pick up a towel from a table by going through a sequence of vision-based re-grasps and manipulations-partially in the air, partially on the table. Even when sufficient manipulation skills are available these need to be constantly adapted by an perception-action loop to yield desired results. 'Robotics, Vision and Control' [16] puts this close integration of the mentioned components into the spotlight and describes common pitfalls and issues when trying to build such systems with high levels of sensorimotor integration. In the DARPA ARM project, which aims to create highly autonomous manipulators capable of serving multiple purposes across a wide variety of applications, the winning team showed an end-to-end system that allows the robot to grasp and pick-up diverse objects (e.g. a power drill, keys, screwdrivers, etc.) from a table by combining touch and LASER sensing [26].

10.5.1   Closing the Action-Perception Loop

The aim is to generate a pick-and-place operation for the *iCub*. For this, functional motion and vision subsystems are integrated to create a closed action-perception loop. The vision side detects and localises the object continuously, while the motor-side tries to reach for target objects avoiding obstacles at the same time. A grasping of the object is triggered when the hand is near the target. The sensory and motor sides establish quite a few capabilities by themselves, yet to grasp objects successfully while avoiding obstacles they need to work closely together. The continuous tracking of obstacles and the

FIGURE 10.7 'Curiously inspect something small': the 3D position of the hand is constrained, and the task space is its angle with respect to the gaze direction. The resulting map rotates the hand (and any grasped object) in front of the eyes.

target object is required to create a reactive reaching behaviour which adapts in real-time to the changes of the environment.

By creating interfaces between *MoBeE* and *icVision* the robot is able to continuously perform a visual based localisation of the detected objects and propagated this information into the world model. This basic eye-hand coordination allows for an adaptation while executing the reaching behaviour to changing circumstances, improving our robot's autonomy.

10.6   RESULTS

The first experiment shows that the herein presented system is able to reactively move the arm out of harms way when the environment changes. Then it is shown how this system can be used to reactively reach and grasp objects.

10.6.1   Avoiding a Moving Obstacle

Static objects in the environment can be added directly into *MoBeE*'s world model. Once, e.g. the table, is in the model, actions and behaviours are adapted due to computed constraint forces. These forces, $f_i(t)$ in (1), which repel the robot from collisions with the table, governs the actual movement of the robot. This way we are able to send arbitrary motions to our system, while ensuring the safety of our robot (this has recently been shown to provide a good reinforcement signal for learning robot reaching behaviours [46, 20]). The presented system has the same functionality also for arbitrary, non-static objects. After detection in both cameras the object's location is estimated (*icVision*) and propagated to *MoBeE*. The fictional forces are calculated to avoid impeding collisions. Figure 10.8 shows how the localised object is in the way of the arm and the hand.[4] To ensure the safety of the rather fragile fingers, a sphere around the end-effector can be seen. It is red, indicating a possible collision, because the sphere intersects with the object. The same is valid for the lower arm. The forces, calculated at each body part using Jacobians, push the intersecting geometries away from each other, leading to a forcing of the hand (and arm) away from the obstacle. Figure 10.9 shows how the the robot's arm is avoiding a non-stationary obstacle.[5] The arm is 'pushed' aside at the beginning, when the cup is moved close to the arm. It does so until the arm reaches its limit, then the forces cumulate and the end-effector is 'forced' upwards to continue avoiding the obstacle. Without an obstacle the arm starts to settle back into its resting pose q^*.

---

[4]See video: https://www.youtube.com/watch?v=w_qDH5tSe7g
[5]See video: https://www.youtube.com/watch?v=w_qDH5tSe7g

10.6.2 Reaching and Grasping Objects

This next experiment is on a simple reactive pick-and-place routine for the *iCub*. Similarly to the above experiment we are using *MoBeE* to adapt the reaching behaviour while the object is moved. To do this we change the type of the object within the world model from 'obstacle' into 'target'. Due to this change there is no repelling force calculated between the object and the robot parts. In fact we can now use the vector from the end-effector to the target object as a force that drives the hand towards a good grasping position.

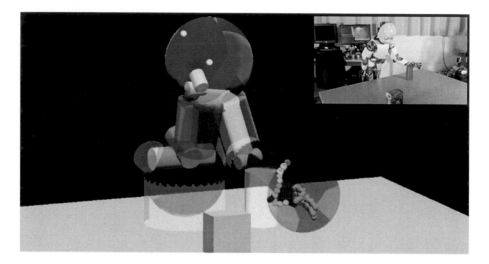

FIGURE 10.8 Showing the visual output of the *MoBeE* world model during one of our experiments. Parts in red indicate (an impeding) collision with the environment (or itself). The inset shows the actual scene.

FIGURE 10.9 The reactive control of the left arm, permitting the iCub to stay clear of the non-static 'obstacle', as well as the table.

MoBeE also allows to trigger certain responses when collisions occur. In the case, when we want the robot to pick-up the object, we can active a grasp subsystem whenever the hand is in the close vicinity of the object. We are using a prototypical power grasp style hand-closing action, which has been used successfully in various demos and videos.[6] Figure 10.6 shows the *iCub* successfully picking up (by adding an extra upwards force) various objects using our grasping subsystem, executing the same action.

Our robot frameworks are able to track multiple objects at the same time, which is also visible in Figure 10.8, where it tracks both the cup and the tea box. By simply changing the type of the object within *MoBeE* the robot reaches for a certain object while avoiding the other.

10.7 CONCLUSIONS

Herein a cognitive robotics approach towards visual guided object manipulation with a humanoid was presented. A tightly integrated sensorimotor system, based on two frameworks developed over the past years, enables the robot to perform a simple pick-and-place task. The robot reaches to detected objects, placed at random positions on a table.

The implementation enables the robot to adapt to changes in the environment. Through this it safeguards the *iCub* from unwanted interactions – i.e. collisions. This is facilitated by a tight integration of the visual system with the motor side. Specifically an attractor dynamic based on the robot's pose and a model of the world. This way a level of eye-hand coordination not previously seen on the *iCub* was achieved.

In the future more integration of machine learning to further improve the object manipulation skills of our robotic system is planned. Improving the predication and selection of actions will lead to a more adaptive, versatile robot. Furthermore it might be of interest to investigate an even tighter sensorimotor coupling, e.g. avoiding translation into operational space by working in vision/configuration space.

10.8 Bibliography

[1] R. Ambrose, B. Wilcox, B. Reed, L. Matthies, D. Lavery, and D. Korsmeyer. NASA's Space Technology Roadmaps (STRs): Robotics, telerobotics, and autonomous systems roadmap. Technical report, National Aeronautics and Space Administration (NASA), 2012.

[2] M. Asada, K. Hosoda, Y. Kuniyoshi, H. Ishiguro, T. Inui, Y. Yoshikawa, M. Ogino, and C. Yoshida. Cognitive developmental robotics: A survey. *IEEE Transactions Autonomous Mental Development*, 1:12–34, 2009.

---

[6]See videos at: http://robotics.idsia.ch/media/

[3] M. Asada, K.F. MacDorman, H. Ishiguro, and Y. Kuniyoshi. Cognitive developmental robotics as a new paradigm for the design of humanoid robots. *Robotics and Autonomous Systems*, 37(2):185–193, 2001.

[4] Tucker Balch and Ronald C Arkin. Behavior-based formation control for multirobot teams. *IEEE Transactions on Robotics and Automation*, 14(6):926–939, 1998.

[5] N.E. Berthier, R.K. Clifton, V. Gullapalli, D.D. McCall, and D.J. Robin. Visual information and object size in the control of reaching. *Journal of Motor Behavior*, 28(3):187–197, 1996.

[6] J. Bongard, V. Zykov, and H. Lipson. Resilient machines through continuous self-modeling. *Science*, 314(5802):1118–1121, 2006.

[7] G. Bradski. The OpenCV Library. *Dr. Dobb's Journal of Software Tools*, 2000.

[8] R.A. Brooks. Intelligence without representation. *Artificial intelligence*, 47(1):139–159, 1991.

[9] R.A. Brooks. *Cambrian intelligence: the early history of the new AI*. The MIT Press, 1999.

[10] E. Brown, N. Rodenberg, J. Amend, A. Mozeika, E. Steltz, M.R. Zakin, H. Lipson, and H.M. Jaeger. Universal robotic gripper based on the jamming of granular material. *Proceedings of the National Academy of Sciences (PNAS)*, 107(44):18809–18814, 2010.

[11] A. Cangelosi, T. Belpaeme, G. Sandini, G. Metta, L. Fadiga, G. Sagerer, K. Rohlfing, B. Wrede, S. Nolfi, D. Parisi, C. Nehaniv, K. Dautenhahn, J. Saunders, K. Fischer, J. Tani, and D. Roy. The ITALK project: Integration and transfer of action and language knowledge in robots. In *Proceedings of the International Conference on Human Robot Interaction (HRI)*, 2008.

[12] A. Chella, M. Frixione, and S. Gaglio. A cognitive architecture for robot self-consciousness. *Artificial Intelligence in Medicine*, 44(2):147–154, 2008.

[13] N.G. Cheng, M.B. Lobovsky, S.J. Keating, A.M. Setapen, K.I. Gero, A.E. Hosoi, and K.D. Iagnemma. Design and analysis of a robust, low-cost, highly articulated manipulator enabled by jamming of granular media. In *Proceedings of the International Conference of Robotics and Automation (ICRA)*, pages 4328–4333, 2012.

[14] Peter C. Chu. SMART Underwater Robot (SUR) Application & Mining, 2011. Technical Presentation.

[15] D. Cliff, P. Husbands, and I. Harvey. Explorations in evolutionary robotics. *Adaptive Behavior*, 2(1):73–110, 1993.

[16] P.I. Corke. *Robotics, Vision and Control*, volume 73 of *Springer Tracts in Advanced Robotics*. Springer, 2011.

[17] B. Dachwald. Optimization of interplanetary solar sailcraft trajectories using evolutionary neurocontrol. *Journal of Guidance, Control, and Dynamics*, 27(1):66–72, 2004.

[18] H. Forssberg, A.C. Eliasson, H. Kinoshita, R.S. Johansson, and G. Westling. Development of human precision grip i: basic coordination of force. *Experimental Brain Research*, 85(2):451–457, 1991.

[19] M. Frank, J. Leitner, M. Stollenga, S. Harding, A. Förster, and J. Schmidhuber. The modular behavioral environment for humanoids and other robots (MoBeE). In *Proceedings of the International Conference on Informatics in Control, Automation & Robotics (ICINCO)*, 2012.

[20] Mikhail Frank, Jürgen Leitner, Marijn Stollenga, Alexander Förster, and Jürgen Schmidhuber. Curiosity driven reinforcement learning for motion planning on humanoids. *Frontiers in Neurorobotics*, 7(25), 2014.

[21] Rafael C. Gonzalez and Richard E. Woods. *Digital Image Processing*. Prentice-Hall, 3rd edition, 2006.

[22] Simon Harding, Jürgen Leitner, and Jürgen Schmidhuber. Cartesian genetic programming for image processing. In Rick Riolo, Ekaterina Vladislavleva, Marylyn D Ritchie, and Jason H. Moore, editors, *Genetic Programming Theory and Practice X*, Genetic and Evolutionary Computation, pages 31–44. Springer New York, 2013.

[23] I. Harvey, P. Husbands, D. Cliff, A. Thompson, and N. Jakobi. Evolutionary robotics: the Sussex approach. *Robotics and Autonomous Systems*, 20(2):205–224, 1997.

[24] B.K. Horn. *Robot Vision*. MIT Press, 1986.

[25] A. Hornberg. *Handbook of machine vision*. Wiley, 2007.

[26] N. Hudson, T. Howard, J. Ma, A. Jain, M. Bajracharya, S. Myint, C. Kuo, L. Matthies, P. Backes, P. Hebert, T. Fuchs, and J. Burdick. End-to-end dexterous manipulation with deliberate interactive estimation. In *Proceedings of the International Conference on Robotics and Automation (ICRA)*, 2012.

[27] M. Jeannerod. *The cognitive neuroscience of action*. Blackwell Publishing, 1997.

[28] Mark H Johnson and Yuko Munakata. Processes of change in brain and cognitive development. *Trends in cognitive sciences*, 9(3):152–158, 2005.

[29] C.C. Kemp, A. Edsinger, and E. Torres-Jara. Challenges for robot manipulation in human environments [grand challenges of robotics]. *IEEE Robotics & Automation Magazine*, 14(1):20–29, 2007.

[30] D. Kragic and M. Vincze. Vision for robotics. *Foundations and Trends in Robotics*, 1(1):1–78, 2009.

[31] J. Leitner, C. Ampatzis, and D. Izzo. Evolving ANNs for spacecraft rendezvous and docking. In *Proceedings of the International Symposium on Artificial Intelligence, Robotics and Automation in Space (i-SAIRAS)*, 2010.

[32] J. Leitner, P. Chandrashekhariah, S. Harding, M. Frank, G. Spina, A. Forster, J. Triesch, and J. Schmidhuber. Autonomous learning of robust visual object detection and identification on a humanoid. In *Proceedings of the International Conference on Development and Learning and Epigenetic Robotics (ICDL)*, November 2012.

[33] J. Leitner, S. Harding, M. Frank, A. Förster, and J. Schmidhuber. icVision: A Modular Vision System for Cognitive Robotics Research. In *Proceedings of the International Conference on Cognitive Systems (CogSys)*, February 2012.

[34] J. Leitner, S. Harding, M. Frank, A. Förster, and J. Schmidhuber. Learning spatial object localization from vision on a humanoid robot. *International Journal of Advanced Robotic Systems (ARS)*, 9, 2012.

[35] J. Leitner, S. Harding, M. Frank, A. Förster, and J. Schmidhuber. Transferring spatial perception between robots operating in a shared workspace. In *Proceedings of the International Conference on Intelligent Robots and Systems (IROS)*, October 2012.

[36] Jürgen Leitner, Alexander Förster, and Jürgen Schmidhuber. Improving robot vision models for object detection through interaction. In *2014 International Joint Conference on Neural Networks, IJCNN 2014, Beijing, China, July 6-11, 2014*, pages 3355–3362. IEEE, 2014.

[37] M. Lungarella, G. Metta, R. Pfeifer, and G. Sandini. Developmental robotics: a survey. *Connection Science*, 15(4):151–190, 2003.

[38] J. Maitin-Shepard, M. Cusumano-Towner, J. Lei, and P. Abbeel. Cloth grasp point detection based on multiple-view geometric cues with application to robotic towel folding. In *Proceedings of the International Conference on Robotics and Automation (ICRA)*, pages 2308–2315, 2010.

[39] M.E. McCarty, R.K. Clifton, D.H. Ashmead, P. Lee, and N. Goubet. How infants use vision for grasping objects. *Child development*, 72(4):973–987, 2001.

[40] A.N. Meltzoff. Infant imitation after a 1-week delay: Long-term memory for novel acts and multiple stimuli. *Developmental Psychology*, 24(4):470, 1988.

[41] G. Metta, P. Fitzpatrick, and L. Natale. YARP: Yet Another Robot Platform. *International Journal of Advanced Robotics Systems, Special Issue on Software Development and Integration in Robotics*, 3(1), 2006.

[42] Giorgio Metta, Lorenzo Natale, Francesco Nori, Giulio Sandini, David Vernon, Luciano Fadiga, Claes von Hofsten, Kerstin Rosander, Manuel Lopes, JosÃl' Santos-Victor, Alexandre Bernardino, and Luis Montesano. The iCub humanoid robot: An open-systems platform for research in cognitive development. *Neural Networks*, 23(8-9):1125–1134, October 2010.

[43] S. Nolfi and D. Floreano. *Evolutionary robotics: The biology, intelligence, and technology of self-organizing machines*. The MIT Press, 2000.

[44] S. Nolfi, D. Floreano, O. Miglino, and F. Mondada. How to evolve autonomous robots: Different approaches in evolutionary robotics. In *Artificial Life IV*, pages 190–197. MIT Press, 1994.

[45] E. Oztop, N.S. Bradley, and M.A. Arbib. Infant grasp learning: a computational model. *Experimental Brain Research*, 158(4):480–503, 2004.

[46] Shashank Pathak, Luca Pulina, Giorgio Metta, and Armando Tacchella. Ensuring safety of policies learned by reinforcement: Reaching objects in the presence of obstacles with the icub. In *Proceedings of the International Conference on Intelligent Robots and Systems (IROS)*, 2013.

[47] J. Peters, S. Vijayakumar, and S. Schaal. Reinforcement learning for humanoid robotics. In *Proceedings of the International Conference on Humanoid Robots*, 2003.

[48] R. Pfeifer, J. Bongard, and S. Grand. *How the body shapes the way we think: a new view of intelligence*. The MIT Press, 2007.

[49] M.I. Posner. *Foundations of cognitive science*. The MIT Press, 1989.

[50] Stuart Jonathan Russell and Peter Norvig. *Artificial Intelligence: A Modern Approach*. Prentice Hall, 3rd edition, 2010.

[51] A. Sadikov, M. Možina, M. Guid, J. Krivec, and I. Bratko. Automated chess tutor. In H.J. Herik, P. Ciancarini, and H.H.L.M. Donkers, editors, *Computers and Games*, volume 4630 of *Lecture Notes in Computer Science*, pages 13–25. Springer Berlin Heidelberg, 2007.

[52] A. Saxena, J. Driemeyer, and A.Y. Ng. Robotic grasping of novel objects using vision. *The International Journal of Robotics Research*, 27(2):157, 2008.

[53] S. Schaal. Is imitation learning the route to humanoid robots? *Trends in cognitive sciences*, 3(6):233–242, 1999.

[54] M. Shanahan. A cognitive architecture that combines internal simulation with a global workspace. *Consciousness and Cognition*, 15(2):433–449, 2006.

[55] M. Stollenga, L. Pape, M. Frank, J. Leitner, A. Förster, and J. Schmidhuber. Task-relevant roadmaps: A framework for humanoid motion planning. In *Proceedings of the International Conference on Intelligent Robots and Systems (IROS)*, 2013.

[56] R.S. Sutton and A.G. Barto. *Reinforcement learning: An introduction*. Cambridge Univ Press, 1998.

[57] N.G. Tsagarakis, G. Metta, G. Sandini, D. Vernon, R. Beira, F. Becchi, L. Righetti, J. Santos-Victor, A.J. Ijspeert, M.C. Carrozza, and D.G. Caldwell. iCub: the design and realization of an open humanoid platform for cognitive and neuroscience research. *Advanced Robotics*, 21:1151–1175, 2007.

[58] D. Vernon, G. Metta, and G. Sandini. A survey of artificial cognitive systems: Implications for the autonomous development of mental capabilities in computational agents. *IEEE Transactions on Evolutionary Computation*, 11(2):151–180, 2007.

[59] D. Vernon, G. Metta, and G. Sandini. The iCub Cognitive Architecture: Interactive Development in a Humanoid Robot. In *Proceedings of the International Conference on Development and Learning (ICDL)*, 2007.

[60] WAY. EU Project Consortium: Wearable Interfaces for Hand Function Recovery. http://www.wayproject.eu/, January 2012.

[61] J. Weng. Developmental robotics: Theory and experiments. *International Journal of Humanoid Robotics*, 1(2):199–236, 2004.

[62] D.M. Wolpert, Z. Ghahramani, and J.R. Flanagan. Perspectives and problems in motor learning. *Trends in cognitive sciences*, 5(11):487–494, 2001.

[63] J. Wyatt and N. Hawes. Multiple workspaces as an architecture for cognition. In A.V. Samsonovich, editor, *AAAI Fall Symposium on Biologically Inspired Cognitive Architectures*. The AAAI Press, 2008.

Index